MATHEMATICS
In Nineteenth-Century
America
THE BOWDITCH GENERATION

TODD TIMMONS

Docent Press

Docent Press
Boston, Massachusetts, USA
www.docentpress.com

Docent Press publishes books in the history of mathematics and computing about interesting people and intriguing ideas. The histories are told at many levels of detail and depth that can be explored at leisure by the general reader.

Cover design by Brenda Riddell, Graphic Details.

Cover photograph of Bowditch by permission of the Boston Athenæum.

Produced with TeX. Textbody set in Garamond with titles and captions in Bernhard Modern.

© Todd Timmons 2013

All rights reserved. No part of this book may be reproduced or utilized in any form or by any means, electronic or mechanical, including photocopying and recording, or by any information storage and retrieval system, without permission in writing from the author.

Contents

Introduction		1
1	A Foundational Period	9
	1.1 One Hundred Years of American Mathematics	10
	1.2 The Bowditch Generation	11
	1.3 The Peirce Generation	15
	1.4 American Mathematics in Context	18
	1.5 Precursor to Community	20
2	The Post-Colonial Generation	25
	2.1 The Role of Colleges in the Development of American Science	36
	2.2 Prospects for American Science	39
	2.3 Physical Science versus Natural History	45
	2.4 David Rittenhouse	47
	2.5 The Changing Face of American Science	52
	2.6 Mathematics in America	55
3	Prosopographical Analysis	59
	3.1 Survey of Publications	62
	3.2 Major Figures in Early American Mathematics	68
	3.3 Mixed Mathematics in American Journals	76
	3.4 Pure Mathematics in American Journals	78
4	Nathaniel Bowditch	87
	4.1 Early Life	93
	4.2 *The New American Practical Navigator*	94
	4.3 Bowditch and the American Academy of Arts and Sciences	102
	4.4 Translation of Laplace's *Mécanique Céleste*	112

5	Making Connections	127
	5.1 The *Transactions of the American Philosophical Society*	130
	5.2 The *Memoirs of the American Academy of Arts and Sciences*	136
	5.3 *The American Journal of Science and Arts*	141
	5.4 The Calculus Question: Differentials or Fluxions, Synthesis or Analysis?	151
	5.5 Early Mathematical Journals	161
6	The French Connection	181
	6.1 French Influence on American Science	182
	6.2 British versus French: Conflicting Mathematical Styles	185
	6.3 The Influence of Jeremiah Day	189
	6.4 Pioneer of the Translation Movement: John Farrar	195
	6.5 Charles Davies and West Point: Emulating the École Polytechnique	212
	6.6 Contributing Factors to the Adoption of a New System of Mathematics	214
Bibliography		217
A	Mathematical Content of American Journals	231
B	Contributions to American Mathematics Journals	233
	B.1 *American Journal of Science and Arts*	233
	B.2 *Transactions of the American Philosophical Society*	237
	B.3 *Memoirs of the American Academy of Arts and Sciences*	240
C	Mathematicians in American Science Journals	243
D	John Farrar's *Cambridge Course of Mathematics*	271

List of Figures

1.3.1	Benjamin Peirce	16
2.4.1	David Rittenhouse	49
3.1.1	Theodore Strong	66
3.2.1	Geometry Problem	70
4.0.1	Nathaniel Bowditch	89
4.2.1	Title page from Bowditch's *New American Practical Navigator* (1802)	97
4.2.2	Cover of the Bicentennial Edition of Bowditch's *American Practical Navigator* (2002)	98
5.3.1	Benjamin Silliman	143
5.5.1	Title page of the first edition of *The Mathematical Correspondent* (1804)	164
5.5.2	Robert Adrain	169
5.5.3	Title page of the first edition of *The Mathematical Diary* (1825)	177
6.3.1	Jeremiah Day	192
6.4.1	John Farrar	197

Acknowledgements

Originating from a Ph.D. dissertation (University of Oklahoma, History of Science Department, 2002), this book is the result of the help and guidance of many people. I would like to express my gratitude to my original dissertation committee, especially Professor Jamil Ragep (chair) for his guidance and advice, and Professor Karen Hunger Parshall, whose expertise in the history of American mathematics proved invaluable throughout the process of researching and writing. I would also like to express my appreciation to all the librarians at the University of Arkansas-Fort Smith, especially Margie Hicks, Martha Coleman, and Carolyn Filippelli, for their tremendous efforts to fulfill my every request. I am also grateful to the Linda Hall Library and its staff (especially History of Science librarians Bruce Bradley and Cindy Rogers) for the opportunity to work—on two different occasions—under a visiting scholar grant in their beautiful facility. Finally, I would like to thank my family: my wife, Becky, for her support (and just as importantly for being a fine editor) and to my children, Michelle and Terry.

Introduction

The first third of the nineteenth century was an important period for the development of American mathematics: Nathaniel Bowditch emerged as a new leader with an international reputation; general topic scientific journals filled a void by publishing mathematical papers until permanent mathematical journals were established later in the century; and American mathematicians began to turn away from the British-dominated mathematical philosophy of their past and to turn towards the modern mathematical approach as represented by the French textbook authors. Each of these factors contributed to a work-in-progress as American mathematicians struggled to build a foundation upon which a research community would form.

This book traces the four components that helped lay that foundation. With the process of building a new nation well underway, this generation of mathematical practitioners was able to turn its attention to the job of laying the groundwork for American science and mathematics. After the turn of the century, a foundation for communicating mathematical ideas began to emerge in the form of a nascent publication community. This publication community, along with an increasing number of professional positions and the introduction of modern mathematical techniques in the American colleges, signified the beginning of the growth towards the critical mass required for the development of a true research community of professional mathematicians. Although this critical mass would not be achieved for some years to come, an increasing number of interested mathematical practitioners signaled the beginning of the long building process. This building process was hastened by the emergence of a leader in the protocommunity, Nathaniel Bowditch.

Bowditch's influence on American science in general and mathematics in particular was immense. His first major work, *The New American Practical Navigator,* firmly established his roots in the American tradition of practical science. The work of the mature Bowditch, however, reflects a new approach, ascribing increased importance to theoretical science. His translation of and commentary on Laplace's *Mécanique Céleste* represents a high water mark for theoretical mathematics in early America. Bowditch's work was acclaimed throughout Europe, but more importantly was offered as an example of the potential of native scholarship in America.

The practical work of Bowditch's youth compared to the theoretical work of his later years is analogous to the changing face of American science and mathematics during the nineteenth century. At the close of the eighteenth century, David Rittenhouse was acclaimed as the premier American mathematician. Rittenhouse's talents and work, however important, would not have established him as a mathematician in the modern sense. Instead, Rittenhouse was a practitioner: a man of many talents whose ability to use instruments and practical mathematics allowed him to stand out among his fellow patriots. By the end of the nineteenth century America could claim true research mathematicians, such as E.H. Moore.[1] Nathaniel Bowditch was a transitional figure who combined the practical bent of the early American republic with the theoretical talents representative of a new century in mathematics.

---

[1] See Karen Hunger Parshall and David E. Rowe, *The Emergence of the American Mathematical Research Community, 1876–1900: J.J. Sylvester, Felix Klein, and E.H. Moore* (Providence, RI: The American Mathematical Society, 1994).

Bowditch also participated in another important factor in the development of American mathematics—an increased awareness for the need to communicate to like-minded practitioners. At the end of the eighteenth century, the first periodicals dedicated to American science emerged. These periodicals, most importantly the *Transactions of the American Philosophical Society* and the *Memoirs of the American Academy of Arts and Sciences* (and later the *American Journal of Science and Arts*), were the first attempts to provide American scientists an outlet for their scientific creativity. Chapter 4 traces the intimate connection between Bowditch and the American Academy of Arts and Sciences.

If the publications of scientific societies represent the initial offerings of a new country, then the first specialized journals symbolize a realized need to shift the center of attention to a narrower and more focused audience. However, no substantial transitions are easily made. Attempts to establish journals devoted entirely to American mathematics failed until late in the nineteenth century. These early failures were primarily due to a lack of a critical mass of interested subscribers. Yet the influence of men such as Robert Adrain represents the first attempts to lay the groundwork for such a publication community.

The first two criteria, establishing a critical mass and forming professional journals for communication, also tie directly to the third. An increase in the number of scientific jobs led to better financial support; however, American mathematics remained many years away from establishing research positions.

The last prerequisite for the development of American mathematics was education. Emerging from the shadow of Britain both politically and scientifically was important for the new nation. Bowditch once again represents a first step in this direction. His ability to read and comprehend the difficult work of Laplace in itself symbolizes an important breakthrough for American mathematics. Bowditch was acclaimed in Europe and in American for making the advanced continental mathematics found in *Mécanique Céleste* accessible to the English-speaking students.

At about the time Bowditch was making his epic translation, John Farrar was initiating his own efforts to bring continental mathematics to American students. Farrar's translations of French classics in mathematics were groundbreaking in themselves. Beginning at Harvard, then at other colleges in America, students were for the first time exposed to a style of mathematics other than that favored in Britain. Farrar regularly chose outdated texts that were soon superseded by the translations of Charles Davies and others, but his work represents the first steps in the transition in American mathematics from a British-dominated mathematical style to a more modern approach with a French influence. This transition, occurring in Britain around the same time, would be a requirement for America to compete with other nations in the fields of modern mathematics.

Although important starting points, the work of a generation of American mathematicians in the first third of the nineteenth century did not immediately result in the establishment of a community of research mathematicians. Instead, it would be another generation to come before such a community emerged. This leaves a gap in the scholarship treating American mathematics in the nineteenth century. A generation of American mathematicians, covering roughly the middle third of the century, built upon the foundations laid by their predecessors and created an atmosphere conducive to the emergence of a modern research community. This generation is traditionally represented by one of the most important figures in nineteenth-century American science, Benjamin Peirce.

Many were influenced by the work of Nathaniel Bowditch, the teaching and texts of John Farrar, and the journals of Robert Adrain, but perhaps none was indebted to the same extent as Benjamin Peirce. Peirce attended grammar school in Salem with Henry Ingersoll Bowditch, Nathaniel's son. While at Harvard, Peirce was taught by Farrar and assisted the elder Bowditch in editing his translation of *Mécanique Céleste*.[2] In the dedication to one of Benjamin Peirce's most important works, *Analytical Mechanics*, Peirce wrote:

> To the cherished and revered memory of my master in science, Nathaniel Bowditch, the father of American geometry.

As a young man, Peirce published his first mathematical work in one of the journals founded by Robert Adrain, the *Mathematical Diary*. Coolidge says of Peirce:

> Here we have a man in quite a different class from any of his predecessors, and able to bear comparison with any who have succeeded him. He had a brilliant and very rapid mind, a profound interest in science, and a conception of a university as a place where mathematics should not only be pursued but advanced, both by teachers and pupils, which was quite different from anything in evidence in America before his time.[3]

In addition to their other contributions to American mathematics and science, the influence of Bowditch, Farrar, and Adrain on Benjamin Peirce made these men crucial to the development of American mathematics.

Although more research is needed to connect the generation of Bowditch, Farrar, and Adrain to that of E.H. Moore, the work done

---

[2] In fact, many years later (1848) Peirce continued to teach Bowditch's translation and commentary of Laplace as professor of mathematics and astronomy at Harvard.

[3] J.L. Coolidge, "Three Hundred Years of Mathematics at Harvard," *American Mathematical Monthly*, 1943, 50:347–356, on p. 350.

during the first third of the nineteenth century must be considered as an important foundation for the development of American mathematics. This work begins the process of establishing an American reputation in theoretical mathematics, as well as the educational and communication networks necessary for American mathematics to emerge as community of professional researchers. For all of these reasons, the period 1800–1838 must be considered as a crucial phase in the history of American mathematics.

In order to accomplish the goal of analyzing the contributions of the Bowditch generation, Chapter 1 begins with a more nuanced survey of the first one hundred years of American mathematics, mathematical practitioners, and the historical literature of the period. We then proceed to an overview that sets the stage for nineteenth-century American science. This overview (Chapter 2) will take a look at the position of American mathematics in the context of the broader American scientific and intellectual communities before 1800. In Chapter 3, an analysis of a developing mathematics publication community emphasizes the mathematics found in three major scientific publications, *The Transactions of the American Philosophical Society*, *The Memoirs of the American Academy of Arts and Sciences,* and *The American Journal of Science and Arts.*

Chapter 4 addresses the influence of Nathaniel Bowditch upon early American science in general and mathematics in particular. Bowditch's major works, *The New American Practical Navigator,* and his translation of and commentary on Laplace's *Mécanique Céleste,* established his reputation among Europeans as a top-notch mathematician and in the process brought new respect to American science and mathematics. Bowditch's work also represents a transition from a culture totally engrossed in the utility of practical science to one that could appreciate the importance of a highly theoretical work such as the *Mécanique Céleste.*

Chapter 5 focuses on the second criterion for a research community, the ability to exchange ideas easily. At first, Americans published their mathematics in general science periodicals such as the *Transactions of the American Philosophical Society* and the *Memoirs of the American Academy of Arts and Sciences*. During the early years of the nineteenth century, various Americans (particularly Robert Adrain) made tentative efforts to establish journals exclusively dedicated to mathematics. Although all such attempts ended in failure, the effort itself reveals that a small but growing number of practitioners were exhibiting interest in such types of communication.

Finally, Chapter 6 examines the changing way in which mathematics was taught at American colleges. This chapter focuses on the evolution of American mathematics from older forms of British-style mathematics to the French-dominated style practiced by most European mathematicians. This was a necessary step in order to improve the quality of mathematics education in the United States and to begin to form the basis of a critical mass of able practitioners.

Chapter 1

A Foundational Period in American Mathematics

In his address to the American Mathematical Society in 1905, entitled "Mathematical Progress in America," society president and co-founder Thomas S. Fiske maintained:

> Before the founding of Johns Hopkins University [in 1876] there was almost no attempt made to prosecute or even to stimulate in a systematic manner research in the field of pure mathematics. Such mathematical journals as were published were scientifically of little importance and as a rule lived but a year or two.[1]

Fiske continued by allowing that there were a few men who deserved mention for their contributions to early American mathematics. In spite of the fact that these men were "for the most part self-trained," they were

---

[1] Thomas S. Fiske, *Bulletin of the American Mathematical Society*, 1905, *11*:238–247, on p. 238.

eminent among their fellows for their mathematical scholarship, their influence upon the younger men with whom they came into contact, and their capacity for research.[2]

These men of stature in American mathematics mentioned by Fiske were Robert Adrain, Nathaniel Bowditch, and Benjamin Peirce. With that, Fiske dismissed the remainder of American mathematics until the time of the arrival of J.J. Sylvester at Johns Hopkins in 1876.

Was this summary dismissal of 100 years of American mathematics justified? If so, from where did the community of American mathematicians emerge after 1876? What role did Adrain, Bowditch and others play in setting the stage for the development of a research community in American mathematics?[3] These are a few of the questions addressed in the following pages.

1.1 One Hundred Years of American Mathematics

Parshall and Rowe call 1776–1876 the "first period" in the history of American mathematics. This 100-year period spans, however, several generations of American mathematical practitioners.[4] Parshall and Rowe see this period as a time when mathematics was

---

[2] *Ibid.*, p. 239.

[3] There is no doubt that 1876 was a watershed year in American mathematics. For a complete treatment of the development of an American research community, see Karen Hunger Parshall and David E. Rowe, *The Emergence of the American Mathematical Research Community 1876–1900: J.J. Sylvester, Felix Klein, and E.H. Moore* (Providence, RI: The American Mathematical Society, 1994).

[4] I use the term mathematical practitioner in an attempt to avoid the implications that exist in the modern definition of the word "mathematician." The practitioners of this generation worked as surveyors, actuaries, and teachers of mathematics, but were certainly not research mathematicians in the modern sense of the word. Although I will use the term mathematician to describe many of these men (and they were very often called mathematicians by their contemporaries), it will be a different use of the word than is commonly employed today.

embedded in the context of general science. American science itself was developing in the college curricula, with the creation of new scientific societies and publications and in seeking more support from the Federal government.[5]

Although this periodization for American mathematics is sound, it does merit closer attention. The period 1776–1876 may be analyzed further in terms of three generations: the "post-colonial" generation, roughly 1776–1800, was characterized by its emphasis on nation-building and political struggle with precious little time left for science and mathematics; the "Bowditch" generation (1800–1838), was characterized by the first attempts by Americans to create a specialized field through education reform and mathematics publications; and the "Peirce" generation, which was characterized by further strides in professionalization through new institutions, publications, and nascent forays into specialized graduate education.

Looking closer at the middle of these three generations allows us to analyze more rigorously the finer structure of the period and to understand better the obstacles Americans faced in building a mathematical community. The mathematical practitioners of the period overcame a few of these obstacles, but many presented barriers not to be overcome until later generations.

1.2 The Bowditch Generation

The first decades after American independence were clearly a time when mathematics, as well as the other sciences, took a backseat to the building of a new nation. Institutions retained much of their colonial character and mathematically-able Americans often suppressed their scientific aspirations for the more immediate concerns

---

[5]Parshall and Rowe, *Emergence*, pp. xiii–xiv.

of the new government. The case of David Rittenhouse, discussed at length in Chapter 2, is a classic example of the sacrifice made by American scientists for the sake of building a nation.[6]

By the turn of the century, however, many of these issues were better under control and some Americans felt they could turn their attentions to more esoteric pursuits such as science and the arts. The presidency of Thomas Jefferson is symbolic of this new zest for scientific work. Jefferson's interest in science, beginning with the founding of the Military Academy at West Point and the scientific charge made to Lewis and Clark, and continuing later with his creation of the University of Virginia, helped to establish a new level of interest and support for science and scientific education.[7] This renewed emphasis placed on education was also reflected in the evolving curriculum of post-colonial colleges.[8]

Smith and Ginsburg call the first half of the nineteenth century "a time of preparation for action"[9] for American mathematics. In a sense, this analysis is true. This period prepared the way for future generations to lead the United States into the realm of research

---

[6] The mathematical landscape of this period has received little attention from historians. Some analysis of mathematics in the post-colonial period may be found in Florian Cajori, *The Teaching and History of Mathematics in the United States* (Washington: Government Printing Office, 1890); John C. Greene, *American Science in the Age of Jefferson* (Ames: The Iowa State University Press, 1984); Stanley M. Guralnick, *Science and the Ante-Bellum American College* (Philadelphia: The American Philosophical Society, 1975); Brooke Hindle, *The Pursuit of Science in Revolutionary America* (Chapel Hill: University of North Carolina Press, 1956); David Eugene Smith and Jekuthiel Ginsburg, *A History of Mathematics in America Before 1900* (New York: Arno Press, 1980); and Dirk J. Struik, *Yankee Science in the Making: Science and Engineering in New England From Colonial Times to the Civil War* (New York: Dover Publications, 1991).

[7] For Jefferson's influence on American science, see Greene, *American Science*. For Jefferson's part in the founding of the University of Virginia, see for instance James Morton Smith, ed., *The Republic of Letters: The Correspondence between Thomas Jefferson and James Madison 1776-1826* (New York: W.W. Norton and Co., 1995).

[8] See especially Guralnick, *Science* and Frederick Rudolph, *Curriculum: A History of the American Undergraduate Course of Study Since 1636* (San Francisco: Jossey-Bass, 1977).

[9] Smith and Ginsburg, *History*, p. 65.

mathematics on par with Europe. However, in another sense this characterization does not do justice to the practitioners of the period. Nathaniel Bowditch, John Farrar, Robert Adrain and other leaders of the small group of interested men were nothing if not men of action. Their work, although not always successful as measured in terms of immediate goals, nevertheless laid a foundation upon which American mathematics would be built.

The beginning of a new century is symbolic of the beginning of substantive changes in American higher education. The first years of the century found specific events contributing to the growth of mathematics in America. The emergence of Nathaniel Bowditch as a leader among American scientists began with his publication of *The New American Practical Navigator* in 1802. Bowditch's importance to American mathematical astronomy is such that Greene marks the publication of his annotated translation of Laplace's *Mécanique Céleste* as the moment that "a door opened through which American astronomers might enter into full participation in the science of celestial mechanics."[10] One goal of the present work is to explore further Bowditch's role in opening this door and his important role in laying the foundation for an American mathematical community.

A second event that signified an embryonic stirring in American mathematics was the creation of the first journal dedicated strictly to mathematics. This journal, founded in 1804 by the English immigrant George Baron, ceased publication after only a few years. The short life span of Baron's journal, as well as several other mathematics journals founded during the Bowditch period, exemplifies Parshall and Rowe's contention that there was not yet a critical mass of practitioners to support such publications. This also speaks to a larger problem in American mathematics: if there did not exist an assemblage large enough to support a single spe-

---

[10]Greene, *American Science*, p. 157.

cialized journal, there were certainly not enough practitioners to found a mathematical *community*.

Nevertheless, a closer look at the short-lived mathematics journals, as well as the other non-specialized journals publishing mathematics during the Bowditch generation, will serve to show how the work of this generation helped to form a foundation upon which future generations might build. A study such as this may shed light on why a mathematical community would not, and could not, form during this period, as well as to illuminate the requirements for such a community to appear. Beginning the present analysis in 1800 thus allows us to focus immediately on those changes under way relative to the development of mathematics in America.

The choice of 1838 as the ending date for this study is first and foremost because this is the year of Nathaniel Bowditch's death. Not only did Bowditch play a central role as a leader of this generation of American scientists, but also the last volume of his influential translation of Laplace's *Mécanique Céleste* appeared in 1839, one year after Bowditch's death.

Two of Bowditch's contemporaries, both of whom played important roles in this era of American mathematics, were also winding down productive careers around the same time. Robert Adrain, whose multiple attempts to establish American mathematics journals is addressed in Chapter 5, resigned from his position as professor of mathematics at the University of Pennsylvania in 1834 and died in 1843. John Farrar, an important textbook writer whose translations from classic French works is discussed in Chapter 6, retired after a long career as Hollis Professor of Mathematics and Natural Philosophy at Harvard in 1836. His series of translations, each of which went through several new editions, all appeared in their last edition within a few years of 1838.[11] Although Farrar

---

[11]In particular, his 7 mathematics texts appeared in their last edition in 1836, 1834, 1837, 1841, 1837, 1840 and 1836.

lived until 1853, his influence on American mathematics was at an end by the time of Bowditch's death.

1.3 The Peirce Generation

As we have seen, by 1838 the careers of three of the most important men in American mathematics had come to an end. It was also about this time that a new talent was just coming to light, Farrar's student and Bowditch's protégé, Benjamin Peirce. Peirce was a professor of mathematics and natural philosophy (later Perkins Professor of Astronomy and Mathematics) at Harvard from 1833–1880. Unlike his predecessors, Peirce came to be known as a research mathematician, the first produced by the United States. Peirce presided over mathematics at Harvard during a period in which significant changes occurred in the mathematical training of college students. Most importantly, it was during Peirce's career that scientific schools were established at Harvard and at Yale, and Peirce's leadership resulted in the first experiments with the elective system at Harvard.[12]

Peirce's rise to influence represents a new era in American mathematics. This, along with the waning influence of the previous generation of Bowditch, Adrain, and Farrar, makes 1838 a logical end to the middle generation of Parshall and Rowe's first period in American mathematics. If 1800–1838 was the "Bowditch" generation in American mathematics, then the period 1838–1876 might be called the "Peirce" generation.

---

[12]For a biographical study of Peirce see Edward R. Hogan, *Of the Human Heart: A Biography of Benjamin Peirce* (Bethlehem: Lehigh University Press, 2008). For Peirce's mathematical work, see I. Bernard Cohen, *Benjamin Peirce: Father of Pure Mathematics in America* (New York: Arno Press, 1980). Also see Raymond Clare Archibald, "Benjamin Peirce," *American Mathematical Monthly*, 1925, *32*:1–30. For his involvement in the general scientific community, also see Robert Bruce, *The Launching of Modern American Science, 1846–1876* (New York: Alfred A. Knopf, 1987).

Figure 1.3.1: Benjamin Peirce

Although not exactly coincidental, the Peirce generation also mirrors closely an era in American science addressed by Robert V. Bruce in his important work, *The Launching of Modern American Science, 1846–1876*. Bruce begins his study in 1846 for several reasons, including the founding of the Smithsonian Institution, the Yale Scientific School, the journal *Scientific American*, and the arrival of Louis Agassiz in the United States.[13] The Peirce generation is essentially identical to the generation studied by Bruce.

While the present work attempts to bring to light answers to questions pertaining to the Bowditch generation, in the process it leaves unanswered other questions concerning the Peirce generation. How did the advent of the elective system influence the study of mathematics in America? Did the elective system play a significant role in the development of a core group of specialists who might be called research mathematicians? What was the role of the new scientific schools at Harvard and Yale in the development of mathematics in America? For instance, did specialized scientific schools necessarily lead to the training of research scientists, especially in mathematics?

A study of the Peirce generation also necessitates a detailed analysis of the mathematics journals established between 1838 and 1876. On the surface, there seems to be a parallel with those journals established in the early part of the century in the sense that none were long-lived. It was not until 1878 that a journal appeared (*The American Journal of Mathematics*) that survives today. Questions arise as to the similarities and differences between the journals of Peirce's period and those of the preceding and later periods.

An auxiliary question pertains to the role played by general scientific journals in the publication of mathematics. Did the *Transactions of the American Philosophical Society*, the *Memoirs of the*

---

[13] Bruce, *Launching*, p. 3.

American Academy of Arts and Sciences, and *The American Journal of Science and Arts* continue to play a part, or did their importance to American mathematics fade? If the latter is true, did other general science journals fill the void?

Finally, questions about the influence of new scientific institutions should be addressed. For instance, what role did the National Academy of Sciences and the American Association for the Advancement of Science, both established in this era, play in the development of American mathematics? Peirce himself was a central figure in the developing American scientific community and his influence on mathematics through these institutions is yet to be established. These are the types of questions that the present study addresses for the Bowditch generation, 1800–1838. Similar questions for the Peirce generation, 1839–1875, would form an important link to the emergence of a research community as described by Parshall and Rowe.

1.4 American Mathematics in Context

In the process of documenting the importance of the period 1800–1838 for the foundations of American mathematics, I also attempt to place mathematics within the larger context of science and other intellectual pursuits in America. As America grew into its independence, literature, painting, and the arts began to take on a distinctively American style. Such authors as Van Wyck Brooks[14] and Russel Blaine Nye[15] have explored this birth of the American intellect.

---

[14]Van Wyck Brooks, *The Flowering of New England (1815–1865)* (New York: Random House, 1936).

[15]Russel Blaine Nye. *The Cultural Life of the New Nation* (New York: Harper and Row, 1960).

American science was at the same time experiencing its own birth pains. This study attempts to demonstrate the interrelationships between mathematics and the broader study of science in America. To do so, the investigations into American mathematics must be situated within the context of colonial and post-colonial science.

In the present work, I strive to show that mathematics, like science and literature, was evolving within the framework of a young and ambitious country. The small group of American mathematical practitioners, still many decades away from forming a professional community, struggled to set their chosen profession on the road towards international respectability. This struggle in the sciences has been well documented by authors such as Greene and Daniels.[16]

Greene, in particular, has done for the general sciences in America what I hope this work accomplishes in a similar way for American mathematics. Greene sets his work in the American context, a context that explores the beginnings of scientific institutions and publications in the United States as well as the new nation's relationship with European science.[17] By exploring emerging institutions and publications supporting the growth of mathematics in America, I will work to show that mathematics, like science in general, is interesting and important for its "difficult beginnings" and "the ultimate consequences of those early strivings."[18]

---

[16]Greene, *American Science* and George H. Daniels, *American Science in the Age of Jackson* (New York: Columbia University Press, 1968). In addition to works previously cited, see I.B. Cohen, *Some Early Tools of American Science* (New York: Russell and Russell, 1967); Alexandra Oleson and Sanborn C. Brown, eds., *The Pursuit of Knowledge in the Early American Republic: American Scientific and Learned Societies from Colonial Times to the Civil War* (Baltimore: Johns Hopkins University Press, 1976); and Nathan Reingold, ed., *The Sciences in the American Context: New Perspectives* (Washington, D.C.: Smithsonian Institution Press, 1979).

[17]Greene, *American Science*, p. 3.

[18]*Ibid.*

Only a few works have addressed mathematics in the early United States. In addition to extensive references made to mathematics in more general science works already mentioned, two particular resources exist for mathematics. Works by Cajori[19] and by Smith and Ginsburg[20] contain a plethora of historical information but are dated and in many ways are found lacking in critical analysis of the historical record. Detailed historical studies of nineteenth-century American mathematics are few. In addition to the landmark study by Parshall and Rowe, works by Ackerberg[21] and Pycior[22] address the subject in some detail and with a high level of critical analysis.

1.5 Precursor to Community

The Bowditch generation represents an important phase in American mathematics. The mathematical practitioners of the first decades of the nineteenth century deserve credit for their influence on Peirce and his generation, as well as for their work in forming many of the structures upon which a community of researchers would be built.

Parshall and Rowe discuss four requirements for the development of a community of scientific researchers: the appearance of a

---

[19] Cajori, *Teaching and History.*

[20] Smith and Ginsburg, *History.* Another valuable resource containing numerous articles concerning American mathematics in the early nineteenth century is Peter Duren, ed., *A Century of Mathematics in America* (Providence, RI: American Mathematical Society, 1988).

[21] Amy Ackerberg, *Mathematics is a Gentleman's Art: Analysis and Synthesis in American College Geometry Teaching, 1790–1840,* unpublished Ph.D. dissertation (Iowa State University, 2000).

[22] Helena Pycior, "British Synthetic vs. French Analytical Styles of Algebra in the Early American Republic," in David Rowe and John Mcleary, eds., *History of Modern Mathematics,* vol. 1 (Boston: Academic Press, 1989), pp. 125–154.

critical mass of practitioners, the ability to exchange ideas easily, financial support for research, and adequate educational opportunities. These requirements were met for American mathematics only in the last quarter of the nineteenth century. However, they did not appear from within a vacuum. The predecessors of J.J. Sylvester, Felix Klein, and E.H. Moore struggled to lay these foundations. Often their struggles ended in failure, but the failures themselves represent first attempts to form a community of like-minded scientists.

This work will document the early attempts by Americans to establish the criteria required for a mathematical community. Although there were an increasing number of men interested in the practice of the mathematical sciences in America, this number was not sufficient to establish a self-perpetuating community of researchers. This failure to form a critical mass of mathematicians may be attributed to several factors—the lack of educational opportunities, an absence of professional positions required to sustain research, a need for established professional journals for publication of research, and a national mindset that did not yet value basic scientific research. Although none of these components existed in the United States, the Bowditch generation of mathematical practitioners began to lay the groundwork for each during the first decades of the nineteenth century.

The establishment of professional journals speaks to the second criterion—the ability to exchange ideas freely. American mathematics authors at the turn of the century relied upon general science journals as avenues to publish their work. It would not be until much later in the century before a professional journal dedicated strictly to mathematics would appear on the scene and survive. However, the first efforts to establish mathematical journals in the United States occurred in the Bowditch era. These journals, several of which were launched by Robert Adrain, were not successful and disappeared after a short time. Their importance lies in the fact that for the first time in America there were at least a *few* men interested in forming a community dedicated to mathematics publication. Furthermore, even during periods in which no specialized journals existed, there was a small but active contingent publishing mathematics in general science journals.

Funding for mathematical research in the form of government and institutional support was yet many years in the future. This is the only one of Parshall and Rowe's four criteria that was not substantially addressed in the period 1800–1838. The few mathematics positions at American colleges required far too much teaching, usually in other areas in addition to mathematics, to allow time for dedicated research. In fact, the leading figure in American mathematics, Nathaniel Bowditch, pursued his mathematical interests while employed by insurance companies in Salem and Boston. Bowditch rejected several offers from American colleges in part because they could not hope to match his financial situation in private business, but also because they could offer little advantage in the pursuit of mathematical interests.

Just as American colleges of the period had little to offer their professors of mathematics besides heavy teaching loads, the colleges had little to offer students in the way of a mathematical education. Following the customs of colonial colleges, students were given a general education that included an introduction to the mathematical sciences but did not offer much hope for advanced study.[23] The Peirce generation would be the first to offer the elective courses of study and the specialized scientific schools needed to form the basis of professional scientific training.

One event of the period 1800–1838 did, however, contribute to this foundation—John Farrar's introduction of French mathemat-

---

[23] For information on the mathematics curriculum in the United States, see especially Cajori, *History and Teaching* and Guralnick, *Science*.

ical methods to the Harvard mathematics curriculum. Farrar's "Cambridge Course of Mathematics," a series of translations made from classic French texts, was a first attempt to bring modern methods to American students. This break with the traditional British mathematics of colonial colleges was an important step in the direction of the modernization of American mathematical education.

Just as the Bowditch generation was important to future generations of American mathematicians, the work done in the postcolonial generation influenced the attitudes and accomplishments of Bowditch and his contemporaries. An outline of the science and mathematics of this period, in the context of a broader American intellectual setting, serves to set the stage for a detailed analysis of the Bowditch generation.

Chapter 2

American Science and Mathematics in the Post-Colonial Generation

For the most part, American science failed to achieve a status equal to European science until well over a century after independence.[1] America's backwardness in science, as well as literature and the arts, has been attributed to many factors. The most common of these factors, however, was the importance the people of the nation attached to the "useful" arts of business and industry. Tocqueville, in his classic work *Democracy in America*, was perhaps the most in-

---

[1] Many books have addressed American science from colonial times through the civil war. See for instance Dirk J. Struik, *Yankee Science in the Making: Science and Engineering in New England from Colonial Times to the Civil War* (New York: Dover Publications, 1991); George Daniels, *American Science in the Age of Jackson* (New York: Columbia University Press, 1968); John C. Greene, *American Science in the Age of Jefferson* (Ames: The Iowa State University Press, 1984); Brooke Hindle, *The Pursuit of Science in Revolutionary America* (Chapel Hill: University of North Carolina Press, 1956); Nathan Reingold, *Science in Nineteenth-Century America* (New York: Hill and Wang, 1964); and Raymond Phineas Stearns, *Science in the British Colonies of America* (Urbana: University of Illinois Press, 1970). For works specific to ante-bellum American mathematics, see Florian Cajori, *The Teaching and History of Mathematics in the United States* (Washington: Government Printing Office, 1890); and D.E. Smith and J. Ginsburg, *A History of Mathematics in America Before 1900* (New York: Arno Press, 1980).

fluential purveyor of this theory. Tocqueville argued that aristocracies favored poetry, painting, and excellent craftsmanship, whereas democracies favored more "practical" pursuits allowing the individual to attain wealth and increased social status.[2] Tocqueville maintained:

> ...few of the civilized nations of our time have made less progress than the United States in the higher sciences or has so few great artists, distinguished poets, or celebrated writers.[3]

He attributed this vacuum to three conditions of American life: the harsh and austere religion of America's founders, the unbounded opportunities for finding fortune in America, and the proximity of Europe, which allowed America to rely on Europeans for its culture.

This analysis of the American condition was a typical view held by Europeans as well as by most Americans in the early nineteenth century. Witness this statement in a popular American literary journal several decades before Tocqueville's work:

> The people of the United States are, perhaps, more distinguished than those of Europe as a people of business; and by an universal attention to the active and lucrative pursuits of life. This habit has grown out of the necessities of their situation, while engaged in the settlement of a new country, in the means of self-preservation, in defending their positions, in removing the obstacles and embarrassments arising from their colonial condition, and in forming and establishing independent systems of government.[4]

---

[2] Alexis de Tocqueville, *Democracy in America* (New York: Harper and Row, 1988), p. 465. This edition is based on Tocqueville's 12th edition (1848), the next-to-last edition published in Tocqueville's lifetime.

[3] *Ibid.*, p. 454.

[4] *The American Review and Literary Journal*, 1801, *1*:iii.

The pursuit of such matters as literature and science was secondary to building a new country and a new way of life.

This new way of life required a new way of thinking about government and its relationship to society:

> There was far too much to know and to say about the rights of man, the nature of government, and the structure of society for men to deal exhaustively with the artistic aspect of life, too much to do to build a state to expend the effort in making a poem.[5]

In addition to the political distractions encountered by anyone who aspired to a life in science, there were the disadvantages to living in a young and vast country. Americans did not have the social and cultural traditions, the rich libraries, the ancient universities, the endowed institutions, or the royal or noble patronage enjoyed by their European counterparts.[6] The sheer size of the country, combined with its geographic isolation from Europe,[7] made communication between population centers, and therefore between scholars, slow and undependable.

In the early nineteenth century, however, there was a growing perception that these pervasive attitudes towards science and the arts were changing:

> When, now, that our population is increased, our national independence secured, and our governments established, and we are relieved from the necessities of colonists

---

[5] Russel Blaine Nye, *The Cultural Life of the New Nation* (New York: Harper and Row, 1960), p. 251.

[6] Hindle, *Pursuit*, p. 3. See also Greene, *American Science*, p. 128.

[7] Isolation from Europe is a common theme among both modern historians and contemporary accounts. Tocqueville's third reason for the paucity of intellectual achievement in the United States, its proximity and reliance on Europe, is a minority view.

and emigrants, there is reason to expect more attention to polite literature and science.[8]

Then the question was, "What was needed to make America an equal contributor to the world's store of knowledge?" The answer seemed evident to turn-of-the-century observers:

> Similar and suitable circumstances [to those found in Europe] would show Americans equally qualified to excel in arts and literature, as the natives of the other continent. But a people much engaged in the labours of agriculture, in a country rude and untouched by the hand of refinement, cannot, with any tolerable facility or success, carry on, at the same time, the operations of imagination, and indulge in the speculations of Raphael, Newton, or Pope.[9]

The author continued by calling for the cultivation of the American intellectual soil. It will take time, the author argued, for America to grow into the intellectual equivalent of Europe.[10] Time, indeed is the only prescription for a country lacking in the sciences as well as the fine arts. John Adams famously wrote to his wife Abigail:

> I must study politicks [sic] and war that my sons may have liberty to study mathematicks [sic] and philosophy.

---

[8] *American Review*, p.iii.

[9] *Ibid.*, p. iv.

[10] This prediction was certainly accurate. In 1826, the New York physician and naturalist James E. Dekay lamented that in Europe, unlike the United States, "Splendid establishments are founded and amply endowed, affording gratuitous instruction in the most minute branches [of science], exhibiting brilliant prospects to the zealous student, and securing to the ripe scholar a secure and honorable retreat in his old age." James E. Dekay, "Anniversary Address on the Progress of the Natural Sciences in the United States: Delivered Before the Lyceum of Natural History, of New-York," reprinted in John C. Burnham, ed., *Science in America* (New York: Holt, Rinehart and Winston, 1971), p. 88. Even half a century later, the American astronomer Simon Newcomb echoed many of the same sentiments regarding the lack of inducements for Americans to pursue careers in science. Simon Newcomb, "Exact Science in America," *North American Review*, 1874, *119*:286–308.

> My sons ought to study mathematicks and philosophy, geography, natural history, naval architecture, navigation, commerce and agriculture, in order to give their children a right to study painting, poetry, musick [sic], architecture, statuary, tapestry and porcelain.[11]

This "first things first" approach exemplified the mindset of many of the intellectual leaders of the young country.

Nineteenth-century commentators perceived many of the same reasons for America's lackluster scientific record as they had for the indifference towards literature and the arts. In 1814, DeWitt Clinton[12] argued that the origins of this indifference were rooted in colonialism. He delineated six specific reasons for the inferiority of eighteenth-century American science:[13]

1. Our pioneers came to this country primarily to acquire wealth, expecting to return to their native lands when this was accomplished. Their loyalty was to their country of origin, not to that of their adoption.

2. They came at a time when the intellectual world was involved in philosophies of words, concerned only with logical subtleties which led to the general neglect of science.

3. The colonial governors showed, in general, no interest in the welfare of the country.

4. That desire for fame which encourages genius was stifled through the lack of an enlightened public.

---

[11] Letter from John Adams to Abigail Adams, post 12 May 1780 [electronic edition]. *Adams Family Papers: An Electronic Archive*. Massachusetts Historical Society. http://www.masshist.org/digitaladams/

[12] Clinton was a political leader in New York who also maintained a correspondence with many scientists and other intellectuals in early nineteenth-century America.

[13] Quoted in Smith and Ginsburg, *History of Mathematics*, p. 6.

5. Racial and religious prejudices rendered impossible any combination of effort to advance science. This is seen in the attitude of the Dutch in New Amsterdam, the French in Canada and Illinois, the Germans in Pennsylvania, and the British in New England.

6. The fact that the parent countries shipped criminals and other undesirable persons to this country prejudiced Europe against the character and possible attainments of our people.

Clinton blamed colonialism and colonial powers for America's failure to achieve intellectual equality. He argued that many Americans desired wealth so they might return to Europe; the government of the colonies was not concerned with the welfare—intellectual or otherwise—of the colonies; and intellectual curiosity was stifled by the uneducated, the immigrants, and the criminals shipped to the new world. All of this mattered little, however, since science of the time was a "philosophy of words" dealing with "logical subtleties." This skewed vision of eighteenth-century science was that of a politician, not a scientist. It also came at a time when the United States was embroiled in a second war with its former colonial masters. In Clinton's view, it was little wonder that America had produced no great thinkers in her early years as she fought to shake off the shackles of colonialism.

Several factors contributed to the realization by Americans that science and mathematics, along with other intellectual pursuits, must become an important part of the new nation. These factors also influenced the role that science and mathematics would play in this national development. Patriotism was one such factor. Nye asserts:

> The more historians, scientists, philosophers, and political theorists studied and speculated about their country, the more they were convinced that it was superior on

almost every count, and that its future greatness was divinely assured.[14]

The call for an "American literature" or an "American science" was a common cause and exerted a tremendous influence on how science developed in America.[15]

Although Americans sometimes exhibited an inferiority complex when it came to European scholarship, patriotism often shone through. Benjamin Franklin was a long-time source of pride for Americans, but he was not as well known as a scientist in America as he was in Europe.[16] Thomas Jefferson's leadership in science, as well as politics, also played an important role in the pride felt by Americans in their own accomplishments and abilities. Jefferson's famous refutation of the French naturalist Count de Buffon is a case in point. Buffon theorized that animals from the old world migrated to the new world where they degenerated due to their surroundings. Jefferson, in his acclaimed work, *Notes on the State of Virginia*, compared the quadrupeds of Europe and America and showed that those living in the new world were in no way inferior to those inhabiting Europe.[17]

---

[14] Nye, *Cultural Life*, p. 44.

[15] Tocqueville, in writing of America's tendency to value its own judgement and discernment, observed, "So, of all the countries in the world, America is the one in which the precepts of Descartes are least studied and best followed." *Democracy*, p. 430.

[16] I. B. Cohen, *Franklin and Newton: An Inquiry into Speculative Newtonian Experimental Science and Franklin's Work in Electricity as an Example thereof* (Philadelphia: The American Philosophical Society, 1956). See especially Chapter 2, "Franklin's Scientific Reputation." See also Joyce E. Chaplin, *The First Scientific American: Benjamin Franklin and the Pursuit of Genius* (New York: Basic Books, 2006).

[17] Jefferson concluded "that of 26 quadrupeds common to both countries [America and Europe], 7 are said to be larger in America, 7 of equal size, and 12 not sufficiently examined." Thomas Jefferson, *Notes on the State of Virginia*, printed in Merrill D. Peterson, ed., *The Portable Thomas Jefferson* (New York: Penguin Books, 1975), p. 88.

The year before the outbreak of the War of 1812,[18] Jefferson wrote John Adams concerning his years of frustration in dealing with the two primary European powers:

> With all their [France and England] preeminence in science... the one is a den of robbers and the other of pirates. And if science produces no better fruits than tyranny, murder, rapine and destitution of national morality, I would rather wish our country to be ignorant, honest and estimable, as our neighboring savages are.[19]

Of course, Jefferson was anything but an anti-intellectual. The man who insisted the Lewis and Clark expedition be treated, at least in part, as a scientific journey, who founded the University of Virginia, and who placed great value on his own broad education was not a man who would choose ignorance. However, Jefferson did dream of a new structure for education based on the Republican principles of his own country.[20]

Another factor influencing the development of science in America was the growing belief that science could be used to make life better for Americans. The utilitarian viewpoint came to dominate the country's perception of science and scientists. Hindle traces this philosophy to the Enlightenment idea of using science for the improvement of society.[21] These ideas were especially important to

---

[18] The war itself enhanced the reputation of the United States, and thus its feelings of patriotism. Europe began to look upon America as more of an equal on the international scene. Donald R. Hickey, *The War of 1812* (Urbana and Chicago: University of Illinois Press, 1989), p. 305.

[19] Alf J. Mapp, Jr., *Thomas Jefferson: Passionate Pilgrim* (New York: Madison Books, 1991), p. 230.

[20] For Jefferson's theories of education, especially as applied to his founding of the University of Virginia, see James Morton Smith, ed., *The Republic of Letters: The Correspondence between Thomas Jefferson and James Madison 1776–1826* (New York: W.W. Norton and Co., 1995).

[21] Hindle, *Pursuit*. See especially Chapter 10.

the young country as it went about the business of building a new society.

This and other Enlightenment ideas, especially as expressed by Scottish moral and natural philosophers, had a particular influence on American science. Many American students traveled to Edinburgh for training and education. Between 1750 and 1790, for example, at least 177 Americans studied at the Edinburgh Medical School.[22] These men, many of whom had a tremendous influence on the development of a nationalistic American science after the Revolution, infused the intellectual themes of the Enlightenment into the thinking of the new nation. Reid-Maroney maintains that these Enlightenment themes, along with a strong dose of Calvinistic practical theology, caused many in the intellectual circle centered at Philadelphia to view the science of government, as well as the science of nature, as an imperfect search for truths that would improve the lot of mankind on earth.[23]

In addition to science and government, Enlightenment ideology influenced the social institutions of the young nation. Always a symbol of brotherhood, post-revolutionary American freemasons began to de-emphasize ancient secret knowledge and instead tried to link freemasonry to "scientific and ingenious men" such as Pythagoras.[24] The masons' interest in education was directly linked to the Enlightenment idea of raising the "vulgar" into a

---

[22] Nina Reid-Maroney, *Philadelphia's Enlightenment, 1740–1800* (Westport, Connecticut: Greenwood Press, 2001), p. 95. See also

[23] *Ibid.*, see especially pp. 162–163. I. Bernard Cohen, *Science and the Founding Fathers: Science in the Political Thought of Thomas Jefferson, Benjamin Franklin, John Adams, and James Madison* (New York: W.W. Norton and Company, 1995) and Timothy Ferris, *The Science of Liberty: Democracy, Reason, and the Laws of Nature* (New York: Harper, 2010)

[24] Steven C. Bullock, *Revolutionary Brotherhood: Freemasonry and the Transformation of the American Social Order, 1730–1840* (Chapel Hill: University of North Carolina Press, 1996), p. 143.

higher state of understanding. Reflecting the Enlightenment idea of the utility of science, John Clark, a nineteenth-century American mason, maintained, "The object for which Masonry is instituted is none other than to make better and happier the human race."[25]

The utilitarian viewpoint favored by most Americans had a long-lasting influence on what sort of science was considered important and therefore would be supported. These utilitarian views were often aimed towards higher education in America. An anonymous author argued:

> ...it will be of infinitely more consequence for you to know the names and uses of vegetable, animal, and mineral productions of your country, than to know the distance and revolutions of all the planets in the solar system. A *few* astronomers are enough for an age, but *every* man should know the history of the substances from which his food—his clothing—his dwelling—his remedies in sickness—and his pleasures in health, are derived.[26]

These preferences for the study of the useful arts had a profound effect on the development of American science for much of its early history.

Interestingly, it was not a foregone conclusion in early American culture that science would inevitably prove useful. Much of natural history could be tied to medicine, but the physical sciences were often perceived to be of questionable utility. American astronomers usually felt compelled to justify their work by pointing out its usefulness to surveying and navigation. Dirk Struik wryly maintains that if the great French mathematician, Joseph Louis Lagrange,

---

[25] *Ibid.*

[26] Anonymous, "A Charge which ought to be delivered to the Graduates in the Arts, in all the Colleges in the United States," in *The Universal Asylum, and Columbian Magazine* (August, 1790). Quoted in Silvio A. Bedini, *Thinkers and Tinkers: the early American men of science* (New York: Scribner, 1975), p. 394.

had immigrated to America, "he would have been able to make a living only as a surveyor."[27] America had not yet developed a perceived need for theoretical science.

Even the use of science in technology and invention was not at all obvious to many. There existed a Swiftonian scorn for science and a related respect for the artisan. Many practical Americans believed industrious men who engineered progress "spent no time in extracting sun-beams from cucumbers."[28] This sort of cynicism towards science stood in the way of Jefferson and other promoters of science in the United States around the turn of the century.

It is apparent, then, that many Americans were aware of their nation's shortcomings in the arts and sciences. However, the mood of the nation may be summarized in the famous quote of Thomas Jefferson, responding to accusations from Europeans that Americans had yet to produce a great poet, scientist, or other thinker. Jefferson pointed to Benjamin Franklin and David Rittenhouse as examples of accomplished American scientists and then concluded by saying:

> When we shall have existed as a people as long as did the Greeks before they produced a Homer, the Romans a Virgil, the French a Racine or a Voltaire, or the English a Shakespeare or a Milton, then there will be cause to inquire the reason.[29]

---

[27] Struik, *Yankee Science*, p. 55.

[28] David Dagget, "Sun-Beams may be Extracted from Cucumbers," (July, 1799) in Burnham, ed., *Science*, p. 47. This is a reference to Swift's famous parody of science. In the same oration, Dagget poked fun at those who believed the future held automatons (horseless carriages), hot air balloons and submarines: "if wood and iron were designed to go alone and carry a load, [then why were] the whole herd of oxen, horses and camels created?" p. 39.

[29] Quoted in Silvio A. Bedini, *Thinkers and Tinkers: the early American men of science* (New York: Scribner, 1975), p. 406.

2.1 The Role of Colleges in the Development of American Science

The status of science in America during the post-colonial period insured that little progress would be made towards building a foundation for a professional community. The dearth of practitioners due to the more important work of building a new nation was only one of the four requirements for the development of a community that was not met during this period. A lack of educational opportunities was also a common problem.

American colleges did not always play a major role in science and mathematics in the eighteenth century. One reason was the relatively insignificant place colleges had in American culture. Frederick Rudolph maintains, "Going to college was one of the least likely things to happen to young men [in America] before 1800."[30] Colleges exerted surprisingly little influence on American science simply because so few attended college. Those young men who did attend college in the eighteenth century were primarily training for the ministry or were young gentlemen with no need of a vocation or career.

Science and mathematics in American colleges progressed very slowly throughout the eighteenth century, sometimes occurring through happenstance. For instance, when in 1714 Yale acquired a collection of over 700 books from England on the subjects of natural philosophy and mathematics, the college was soon prompted to insert algebra into the curriculum (1718) so that students could study the contents of the new library, especially Newtonian physics.[31]

Mathematics at Harvard was patchy at best, although the 1719 commencement theses included "A fluxion is the velocity of an in-

---

[30] Frederick Rudolph, *Curriculum: A History of the American Undergraduate Course of Study Since 1636* (San Francisco: Jossey-Bass, Inc., 1977), p. 25.

[31] *Ibid.*, p. 33.

creasing or diminishing flowing quantity" and "A fluxion is found from a flowing quantity."[32] It would be over a century later (1821) before American education advanced to a point that allowed Columbia to become one of the first American colleges to require algebra for admission.[33]

Even after the turn of the century, science was seldom a central point in the college curriculum. Therefore, a science professor, if a college employed such a person, was often in a precarious position:

> The science professor, to apply the term loosely, was at best a peripheral entity in the collegiate organization of the opening decades of the century. His salary was uniformly lower than that of other professors, and his security such that he easily fitted the classic mold of "last hired and the first fired."[34]

Positions in science and mathematics were few in eighteenth-century America. The first scientific professorship was established at William and Mary in 1711.[35] In 1726, Isaac Greenwood was appointed to the newly created Hollis Chair of Mathematics and Natural Philosophy at Harvard, becoming what Rudolph calls "the country's and Harvard's first scientist."[36] Greenwood had previously spent time in London studying the Newtonian system with

---

[32] I. Bernard Cohen, *Some Early Tools of American Science* (New York: Russell and Russell, 1967), p. 62. Newton's fluxions were the predominant style of calculus in America (after all, still a part of Great Britain) in the eighteenth century. It was not until the second quarter of the nineteenth century that serious advances were made in introducing the continental differential calculus into the curriculum of American colleges.

[33] Stanley M. Guralnick, *Science and the Ante-Bellum American College* (Philadelphia: The American Philosophical Society, 1975), p.34.

[34] *Ibid.*, p. 142.

[35] Rudolph, *Curriculum*, p. 34. See also J.L. Coolidge, "Three Hundred Years of Mathematics at Harvard," *American Mathematical Monthly,* 1943, 50:347–356, on pp. 347–349.

[36] Rudolph, *Curriculum*, p. 34.

Desaguliers.[37] In his position as Hollis professor, Greenwood gave lectures on such topics as algebra, conic sections, and the method of fluxions.[38]

The Hollis chair of mathematics and natural philosophy was the closest approximation in America to a professional mathematics position, although its holder often was not a true mathematician and was expected to lecture, not engage in research. After Isaac Greenwood, John Winthrop, an astronomer of some note, held the chair for nearly half a century. He was succeeded by Samuel Williams in 1779. Williams authored many manuscripts for use in his astronomy, mathematics, and natural philosophy classes. He also authored the first college mathematics textbook written by an American.[39] Samuel Webber, a non-descript mathematician but the future president of Harvard, succeeded Williams in the Hollis chair and held the position at the turn of the century.

A chair of mathematics and natural philosophy was created at Yale in 1770, with Nehemiah Strong as its first professor. It was not until after the turn of the century, however, that a mathematician of some repute, Jeremiah Day, held the chair at Yale. Similarly, the chair of mathematics at the University of Pennsylvania was not created until 1779, although its first occupant, Robert Patterson, was a relatively influential man in early American mathematics.[40]

By 1776, six colleges in America had a professor of mathematics and natural philosophy; by 1788, two others had followed suit.[41] Still, by one count, there were only twenty-one full time positions

---

[37] Hindle, *Pursuit*, p. 82.

[38] Bedini, *Thinkers and Tinkers*, pp. 156–157.

[39] Cajori, *Teaching and History*, p. 52.

[40] For a detailed discussion of the mathematicians at various American colleges, see *Ibid.*

[41] Rudolph, *Curriculum*, p. 35.

in science available in America in 1802.[42] By 1840, the American scientific community numbered somewhere between 300 and 600.[43] Although this increase in numbers does not necessarily speak to the quality of instruction in American colleges or the quality of work produced by American scientists, it does indicate that for the first time America was producing a significant quantity of scientists and science teachers. These numbers would at the very least increase the possibility that research scientists on the level of European scientists could be produced in America. In other words, America was progressing towards a "critical mass" of scientists. Whether it also neared a critical mass of mathematicians is one question the present work strives to answer.

2.2 Prospects for American Science

In spite of the dire position of science in the colleges of the eighteenth century, there was optimism in America that science and other intellectual pursuits would begin to flourish. Interestingly, some of the optimism surrounding the prospects for American science and arts around the turn of the century had existed in much the same fashion half a century earlier in colonial America. In 1743, Benjamin Franklin wrote:

---

[42]Burnham, *Science*, p. 7. Burnham does not elaborate on how this count was made. A similar count by Clark Elliott indicates Burnham's numbers might be too low. Elliott supplies the data for American scientists born before 1776 (and who, logically, would be included in the employment picture in 1802, twenty-six years later). Elliot lists 20 Americans employed as professor of science, 6 in science-related fields in the government and 30 in science-related fields in nongovernment positions. Care must be taken in interpreting these numbers however, as Elliott counts American scientists in up to four employment categories if they held more than one position. Clark A. Elliott, *Biographical Dictionary of American Science: The Seventeenth Through the Nineteenth Centuries* (Westport, CN: Greenwood Press, 1979), p. 5.

[43]Burnham, *Science*, p. 73. By comparison, Elliot's count of Americans born between 1776 and 1815 shows 88 employed as professors of science, 64 in governmental science positions and 60 in nongovernmental science positions. Elliot, *Biographical Dictionary*, p. 5.

> The first drudgery of settling new colonies which confines
> the attention of people to mere necessaries is now pretty
> well over; and there are many in every province in cir-
> cumstances that set them at ease, and afford leisure to
> cultivate the finer arts and improve the common stock of
> knowledge.[44]

Of course, major disruptions to the growth of science in America occurred during the Revolutionary War. The British occupied major cities, colleges closed, libraries were damaged, journals suspended, and men of science were pressed into political or military service, with some Loyalists leaving America after the war.[45] Brooke Hindle maintains the war "killed much of the remaining encouragement and peace of mind that permitted the growth of creative science."[46] Many who looked back at science in America after the war held this position. One nineteenth-century commentator concluded:

> ...in any review of the progress of science,...the period
> which lies between the declaration of independence and
> the close of the eighteenth century may, without danger
> of any important omission, be passed over in silence.[47]

Post-war nationalism provided some stimulus to education. For instance, from 1636 to 1776, only nine colleges were established in

---

[44] Quoted in Hindle, *Pursuit*, p. 1.

[45] Nye, *Cultural Life*, p. 70.

[46] Brooke Hindle, *David Rittenhouse* (Princeton: Princeton University Press, 1964), p. 110. Hindle also points out that war had more of a disruptive influence on the physical sciences than on natural history. The majority of work in physical sciences occurred in the colleges, which experienced major disruptions during the war. Natural history, on the other hand, was done primarily outside the colleges in the countryside, often in places not as affected by war. Hindle, *Pursuit*.

[47] F.A.P. Barnard, *The First Century of the Republic*. Quoted in Smith and Ginsburg, *History*, p.17.

America.[48] From 1782 to 1800, at least sixteen new colleges were established (some with an admittedly short life span).[49] On the negative side, this sudden expansion adversely affected standards, as there were neither enough qualified people to teach nor enough money to pay them.[50]

In spite of the post-war gains in higher education, science continued to struggle to establish a foothold in the new nation. In many ways, the post-war atmosphere simply traded one set of problems for another. Whereas in the earlier part of the eighteenth century Americans were concerned with the geographic and topographic problems of settling a new land, the citizens of the new United States of America were equally distracted by the political requirements of building a new nation. Prospective men of science like Benjamin Franklin, Thomas Jefferson, and David Rittenhouse spent much of their time and energy in constructing a novel political and social archetype for their new country.

Science was not alone in its struggle to maintain a foothold in the early Republic. Individuals interested in other intellectual pursuits found themselves in much the same position. John Trumball, an American painter, lamented that painting in the United States was regarded as "frivolous, little useful to society, and unworthy of a man who had talents for more serious pursuits."[51] Portrait painting, an art form that might be deemed "useful," was the only avenue for financial success in early America.

---

[48]Nye, *Cultural Life*, p. 175–176. These colleges were: Harvard (1636), William and Mary (1693), Yale (1701), College of New Jersey, now Princeton (1746), King's College, now Columbia (1754), College of Philadelphia, now University of Pennsylvania (1755), Brown (1764), Queen's College, now Rutgers (1766) and Dartmouth (1769). H.G. Good, *A History of Western Education* (New York: The Macmillan Company, 1947), pp. 391–392.

[49]Nye, *Cultural Life*, p. 176.

[50]*Ibid.*

[51]Quoted in *Ibid.*, p. 277.

Literature, like art and science, was expected to be useful. It was expected that "novels should instruct, dramas draw moral lessons, satires discover and castigate error, essays debate and argue, poetry please and teach."[52] These attitudes inevitably led to a paucity of artists, writers, and scientists. John Pickering lamented:

> in this country we can hardly be said to have any *authors by profession*. The works we have produced have, for the most part, been written by men, who were obliged to depend upon other employments for their support, and who could devote to literary pursuits those few moments only, which their thirst for learning, stimulated them to snatch from their daily avocations.[53]

Would-be scientists found themselves in the same position in America as did artists, writers and other intellectuals. Fulfillment of the third criterion for a community of professionals, adequate financial support, remained far in the future.

These difficulties encountered by American intellectuals reflected the general differences between European society and the society under development in the United States. Nathan Reingold summarizes these societal differences and their affects on the development of science in America:

> Europeans established a scheme of things in which there was a congruence between social, intellectual, and institutional hierarchies—not a perfect congruence, but enough to avoid many of the problems Americans had. On the western side of the Atlantic, in comparison, the heritage of the past and the thrust of historical development did not neatly separate grand savant and practitioner, the-

---
[52] *Ibid.*, p. 262.

[53] *A Vocabulary or Collection of Words and Phrases* (1816). Quoted in *Ibid.*, pp. 248–249.

oretician and earnest mechanic, the abstruse and the vernacular. They were scrambled together. Thus, a persisting tension developed between the scientific elite, with its perception of fundamental research, and the mass community, with its thrust for diffusion of knowledge (often older and sometimes applied). As a result, the cult of knowledge in the United States encompassed a research ideal but not a basic research ideal, despite all the exertions of many generations of scientists.[54]

It was this setting in which American intellectuals of all types, not only scientists, found themselves in the first decades of the new Republic.

Obviously, obstacles to and even disinterest in science, as well as other intellectual pursuits, marked the early national period in the United States. The first decades of the nineteenth century, however, found a young nation increasingly willing to consider the importance of such activities as science, literature, and the arts.

Many factors combined to heighten interest in science in America in the first half of the nineteenth century. A new patriotism spread throughout the country after the War of 1812.[55] Americans began to resent European superiority in any endeavor and science was no exception. Daniels traces the origins of a true American scientific community to this newfound patriotism, as well as to the growth in journals and scientific societies in the United States.[56]

---

[54]Nathan Reingold, "Reflections on 200 Years of Science in the United States," in Nathan Reingold, ed., *The Sciences in the American Context: New Perspectives* (Washington, D.C.: Smithsonian Institution Press, 1979), p. 18.

[55]See, for instance, Guralnick, *Science* and Daniels, *American Science*.

[56]Daniels, *American Science*, p. 4.

One interesting consequence of this sudden growth in science after the War of 1812 was the astounding increase in the number of scientific societies founded in the United States. In 1785 there were three such societies in the United States.[57] By 1815, this number had only grown to seven. Yet in 1825, there were twenty-three scientific societies in the United States, an increase of over 228% in one decade! These numbers continued to grow into the second half of the nineteenth century, but at a much slower rate. The second and third decades of the century witnessed tremendous growth of interest in science in the United States.

The first years of the nineteenth century also signaled many improvements to the plight of science and mathematics in American colleges. In 1802 two particular happenings proved critical to the future of American science. First was the establishment of the Military Academy at West Point. West Point was the first technical college established in the United States, and its emphasis on mathematics and engineering had a great influence over American science. Also in 1802, Benjamin Silliman was appointed professor of chemistry, geology, and mineralogy at Yale. Silliman's leadership proved vital to the growth of American science throughout the first half of the nineteenth century.[58]

In addition to the development of West Point, the budding career of Benjamin Silliman, and the founding of new scientific societies and new colleges, other factors contributed to the growth of American science in the first decades of the nineteenth century. Agricultural and geological surveys arose and new technologies supported an environment conducive to scientific exploration.[59]

---

[57] The counts compared here come from Ralph Bates, *Scientific Societies in the United States* (Cambridge, MA: The MIT Press, 1965), p. 51.

[58] Silliman's role in promoting mathematics through his journal, *The American Journal of Science and Arts*, is discussed in Chapter 5.

[59] Greene, *American Science*, p. 410.

Furthermore, the leading scientific societies, such as the American Philosophical Society and the American Academy of Arts and Sciences, began to encourage more basic scientific research instead of insisting on contributions to so-called "practical" subjects.[60]

In spite of these encouraging events, science (and other academic disciplines) continued to meet with resistance at every turn. Guralnick argues, "Plotting any new course for the college was made difficult by inherent working-class abhorrence for intellectual expertise, no matter how practical the end for which it might be diverted."[61] Guralnick's argument echoes those made by Tocqueville in *Democracy in America*.

2.3 Physical Science versus Natural History

While America made some scientific inroads in the early years, the physical sciences were usually at a distinct disadvantage to other sciences, especially those that fall under the rubric of natural history.[62] Americans made significant contributions to sciences such as mineralogy, zoology, botany and other life and earth sciences long before American physical scientists were able to stand alongside their European counterparts as equals. There are many reasons for the primacy of the natural sciences over mathematics, physics, chemistry and other physical sciences in early America.

One contributing factor was the technical and mathematical education required for many of the physical sciences. The pursuit of natural history, on the other hand, was often a matter of exploration and classification and thus required less rigid technical

---

[60] *Ibid.*, p. 418.

[61] Guralnick, *Science*, p. 119.

[62] For American contributions to natural history see especially Hindle, *Pursuit*, and Greene, *American Science*.

training. In addition, there existed in America a ready-made group of educated men interested in natural history: physicians. Physicians often had the training and even the wealth and leisure time to engage in collecting and classifying the new species found in their own regions.

Another advantage natural history had over the physical sciences was its "entertainment value." A botanist, for instance, might have an impressive garden to show interested guests with little scientific training. A naturalist or collector often possessed a cabinet of curiosities that induced interest from a variety of people. Charles Wilson Peale's famous museum in Philadelphia was a classic example of natural history sparking interest in the non-scientific public. Mathematics and the physical sciences usually did not have the power to excite the interests of the public in the way natural history could.

Of course, some aspects of the physical sciences did present limited entertainment value. Astronomical observations were easily made, either with the naked eye or through a hand-held telescope. Although of some interest to the public, astronomy struggled to gain a prominent place in American culture, as evidenced by the nearly century-long battle to establish a permanent observatory in the United States.[63]

A celebrated exception to the primacy of natural history was Benjamin Franklin's interest in electricity. Many electrical scientists performed their experiments for an awe-struck audience. Some of these experimenters even invited the audience to participate, as when a large number of people would join hands to test the conductivity of electricity through the human body. Franklin himself planned to entertain a social gathering by using electricity to kill

---

[63] For early American efforts to establish permanent observatories, see Green, *American Science*; Robert Bruce, *The Launching of Modern American Science, 1846–1876* (New York: Alfred A. Knopf, 1987); and Marlana Portolano, "John Quincy Adams' Rhetorical Crusade for Astronomy," *Isis*, 2000, *91*:480–503.

and roast a turkey, kindle a fire and discharge guns.[64] How could mathematics compete with such a show?

The most significant factor contributing to America's participation in the international natural history community was geographical. The New World offered flora and fauna never before known to scientists, and Americans had the advantage of location: they lived among this plethora of new species. Collecting and classifying seemed a natural thing to do for Americans interested in the sciences. The life and work of John Bartram is a case in point. A farmer by occupation, Bartram was able to travel America collecting specimens thanks to the patronage of interested European scientists and supporters of science.[65] With a few exceptions, American physical scientists did not enjoy the same advantage of place as did those working in natural history.[66]

2.4 David Rittenhouse

After Benjamin Franklin, the two best-known physical scientists produced in America before the nineteenth century were John Winthrop and David Rittenhouse. As already mentioned, Winthrop was the second Hollis professor of mathematics and natural philosophy at Harvard, serving in that capacity for forty-one years. He discovered one of the moons of Jupiter[67] and lectured at Harvard on the method of fluxions as early as 1751.[68] He published

---

[64]Reid-Maroney, *Philadelphia's Enlightenment*, p. 21.

[65]Hindle, *Pursuit*, p. 27. Hindle points out, though, that American natural historians were primarily collectors and fieldworkers who sent their findings back to Europe for study and classification. p. 79.

[66]One exception was the observations of the transit of Venus in 1769. The significance of this event is discussed later in the chapter.

[67]Bedini, *Thinkers and Tinkers*, p. 74.

[68]Guralnick, *Science*, p. 11.

eleven papers on astronomy in the *Philosophical Transactions* of the Royal Society[69] and was elected to the Royal Society in 1766. Greene maintains, "Winthrop's performance [as a scientist] was not to be equaled or surpassed for sixty years after his death in 1779."[70]

David Rittenhouse was a self-taught astronomer and mathematician[71] who is best remembered as an instrument maker, having started his career as a clockmaker. He spent much of his life in search of a position that would allow him the leisure time to pursue his scientific studies. Although his friends in Philadelphia attempted to create an official post as "Public Observational Astronomer" specifically for Rittenhouse, he never was able to obtain a position that would allow him the time he needed to pursue his research.[72] At various times Rittenhouse was employed as a surveyor and as a professor of astronomy at the University of Pennsylvania, but neither allowed him the time or opportunity for research.

Rittenhouse published numerous papers in the *Transactions of the American Philosophical Society*. In fact, his papers on the famous orrery he constructed and on his observations of the transit of Venus in 1769 dominated the first volume of the *Transactions*. Rittenhouse succeeded Franklin as president of the American Philosophical Society and was elected as a foreign member of the Royal Society of London.

The life and work of David Rittenhouse may be seen as a microcosm of American science before 1800. Although revered as

---

[69] Hindle, *Pursuit*, p. 87.

[70] Greene, *American Science*, p. 76.

[71] As was discussed in Chapter 1, the term *mathematician* is problematic. At different periods in American history, mathematician meant different things. Silvio Bedini, in his book *Thinkers and Tinkers: Early American Men of Science*, points out that in colonial America a mathematician was often anyone who could use mathematical instruments (p. 63). A surveyor, then, was a mathematician.

[72] Hindle, *Rittenhouse*, p. 41.

Figure 2.4.1: David Rittenhouse

America's leading man of science,[73] Rittenhouse was, throughout his life, much too busy with important matters of life in a new country to achieve his full potential as a scientist. Rittenhouse might have accomplished much more in science if it were not for the Revolutionary War. Throughout the conflict, in the prime of his intellectual life, Rittenhouse's attentions were diverted to helping the war effort, rather than focused on science. After the war, politics intruded on his time, as did practical jobs for the new Republic such as land surveys, canal-building, and a stint as Director of the Mint. Many influential Americans despaired that Rittenhouse was too busy to devote his life to science. Thomas Jefferson, referring to Rittenhouse, wrote:

> Nobody can conceive that nature ever intended to throw away a Newton upon the occupations of the crown ...the commonplace drudgery of governing a single state ...may be executed by men of an ordinary stature, such as are always and every where to be found.[74]

Rittenhouse himself was keenly aware of his latent talents, and what might have been if he had lived in a different time or place. He was said to have told a friend:

> If I were independent in my fortune and free to devote myself to my passion for astronomy and science, I would produce a work which would fill Europe with astonishment.[75]

Unfortunately, the United States was not yet ready to support basic science with public funds and Rittenhouse never achieved

---

[73] Rittenhouse and James Bowdoin were the only Americans elected to the Royal Society between the Revolutionary War and 1800. *Ibid.*, p. 357.

[74] Quoted in *Ibid.*, p. 201.

[75] *Ibid.*, p. 249.

the revolutionary work he thought himself capable of producing. In this sense, we see two of our criteria for the development of a community of professionals bound together. Without adequate financial support, the possibility for the development of a critical mass of practitioners was remote.

Another Rittenhouse quality that was common to men of science in eighteenth-century America was his emphasis on the utility of his own work. Trained as a clockmaker and famous as an instrument maker, Rittenhouse described himself as a "mechanic" interested in the practical applications of science.[76] In fact, Rittenhouse disdained mathematics for its own sake. In response to a set of arithmetic problems given to him by his brother-in-law, Rittenhouse wrote, "You cannot conceive how much I despise this kind of juggle, where no use is proposed."[77] Even when Rittenhouse worked on seemingly theoretical questions, they had a practical basis. For instance, in the third volume of the *Transactions of the American Philosophical Society*, Rittenhouse published a paper that gave a method for summing the powers of the sine function.[78] The paper was not written by Rittenhouse for the sake of pure mathematics, but rather was related to a question in clockmaking.

After Franklin's experiments with electricity, possibly the most important event in the physical sciences in eighteenth-century America was the observation of the transit of Venus in 1769.[79] Americans participated in the observation of this transit from several locations. Their observations and calculations, along with those made from European locales and others across the globe,

---

[76] *Ibid.*, p. 83.

[77] *Ibid.*, p. 89.

[78] John Wallis had discovered his method, unbeknownst to Rittenhouse, a century earlier. *Ibid.*, p. 329.

[79] For an account of this important event, see Hindle, *Pursuit*.

provided important data required for better estimates of the planetary distances. These observations also focused attention on the observational astronomers, like John Winthrop and David Rittenhouse, residing in America. Hindle calls the American contribution to this worldwide study "the major factor in gaining a new recognition for American science."[80]

It is significant that America's most important contributions to astronomy were similar to those that occurred in natural history. They were primarily observational in nature and important because of America's unique location in the world. Observations and calculations leading to a determination of the precise location of American cities were important not only for the geographical concerns of the young country, but also as benchmarks for other observations such as the transit of Venus discussed earlier. Theoretical contributions by American scientists were almost non-existent.

Astronomical observations like those made during the transit of Venus were central to the early volumes of the *Transactions of the American Philosophical Society* and the *Memoirs of the American Academy of Arts and Sciences*. Although not specialized journals, these publications did provide an outlet for Americans beginning to show interest in publishing their work at home.

2.5 The Changing Face of American Science

Historians have applied many generalizations to American science around the turn of the nineteenth century. As we have seen, natural history predominated over the physical sciences, utility and practicality took precedence over theory, little specialization occurred in the sciences, and science was primarily a pursuit of the educated gentleman. The concept of a professional scientist in the modern

---
[80] *Ibid.*, p. 165.

use of the word was unknown. The few Americans who made their living in science were usually employed at the colleges where they spent the vast majority of their time teaching.

These generalizations concerning American science began to change after the first decades of the nineteenth century, the "Jacksonian" period of American history. For instance, George Daniels finds almost as many Americans working in the physical sciences as were working in natural history during this period.[81] This represents a significant shift in emphasis from the early national period when American science was usually synonymous with natural history. Daniels also finds as much work done on the "impractical" sciences of zoology and geology as on the practical sciences of botany and mineralogy.[82] He maintains American scientists of the period were becoming more interested in advancing knowledge, but continued to submit practical reasons for their research to win public, and therefore governmental, support.[83] Furthermore, Daniels argues that specialization in science was quickly becoming the norm, citing evidence that most of the scientists of the period realized the implications of specialization.[84]

One of these implications was that American science was no longer primarily a pursuit for amateurs. Scientists began to receive increasingly higher levels of training, albeit many in medicine. Specialization and the ensuing professionalization of science later in the century meant that the public could no longer share in the knowledge of science.[85] Therefore, it was important that "the emerging

---

[81] Daniels, *American Science,* p. 19. A detailed analysis of the Americans working in mathematics itself is the subject of the next chapter.

[82] *Ibid.*, pp. 20–21.

[83] *Ibid.*, p. 21, 24–28.

[84] *Ibid.*, p. 30.

[85] Professionalization in the modern sense of the word was yet many decades away. A generation of scientists in the mid-nineteenth century, including Benjamin Peirce, Louis Agassiz,

profession...justify its work in terms of its social value."[86] Hence, scientists were forced to frame their work in terms of its usefulness to society, even when theoretical knowledge was the ultimate goal of the investigator.

Although American science in the period 1800–1838 certainly cannot be categorized as equal to its European counterparts, certain steps were taken to close the gap between American and European scientists. A small but growing number of scientists at American colleges were beginning to influence a new generation of students. The physical sciences, which required more technical training in mathematics and other areas, began to emerge on equal footing with natural history. Along with better educational opportunities, the more stable political climate allowed many to choose professions such as science and the arts in which to apply their considerable intellect. A man with the talents of Jefferson or Rittenhouse might now take the opportunity to pursue interests outside of politics and nation-building. Finally, specialization and professionalization were important if America had hopes of keeping up in the fast changing world of science. A few young American scientists realized this and began to concentrate their efforts on attaining expertise in specific branches of science.

Joseph Henry, and Alexander Dallas Bache, recognized the need for the professionalization of American science. It would not be until near the end of the century, however, that a true community of professionals, dedicated to research, would emerge in America. For a study of the professionalization of American mathematics see Karen Hunger Parshall and David E. Rowe, *The Emergence of the American Mathematical Research Community, 1876–1900: J.J. Sylvester, Felix Klein, and E.H. Moore* (Providence, RI: The American Mathematical Society, 1994).

[86]Daniels, *American Science*, p. 41.

2.6 Mathematics in America

Mathematics in America, first in the colonies and later in the young nation, may be characterized as unsophisticated and British-dominated.[87] Although a few men of ability maintained an interest in mathematics, and some such as David Rittenhouse and John Winthrop even gained a fair amount of recognition, little was done in the way of research or advanced training of American students. In *A History of Mathematics in America Before 1900*, Smith and Ginsburg maintain:

> The first three centuries of our history were, as we have seen, barren of achievement in the domain of mathematics. The first half of the nineteenth century was a time of preparation for action.[88]

Brooke Hindle, in *The Pursuit of Science in Revolutionary America*, echoes the sentiments of Smith and Ginsburg:

> Few of the sciences were as sterile as mathematics, for most of them had descriptive phases which the Americans could illuminate even though they were unable to improve upon basic theory.[89]

Hindle also states simply, "No one did anything to advance mathematics [in pre-revolutionary America]."[90]

These opinions expressed by modern historians are no different from the opinions held by contemporary commentators on the state

---

[87] For works on American mathematics in the eighteenth and nineteenth centuries, see Cajori, T*eaching and History*; Peter Duren, ed., *A Century of Mathematics in America*, 3 vols., (Providence, RI: The American Mathematical Society, 1988); Parshall and Rowe, *Emergence*; and Smith and Ginsburg, *History*.

[88] Smith and Ginsburg, *History*, p. 65.

[89] Hindle, *Pursuit*, p. 94.

[90] *Ibid.*, p.188.

of American mathematics. When John Farrar was appointed Hollis Professor of Mathematics and Natural Philosophy at Harvard, his qualifications as a scientist were certainly minimal. His appointment demonstrated the low state of mathematics and physics in America in the early nineteenth century, as reflected in the following obituary notice:

> There were probably not half a dozen persons in New England who knew the rudiments of the differential calculus. Webber's *Course of Mathematics*, beginning with Numeration, and closing with an elementary chapter on Spherical Astronomy, was the College text-book [at Harvard], and unless our memory is greatly at fault, even the end of that manual was never reached or approached by any class. Pure mathematics was a very unwelcome study. Nine tenths of every class had broken down in quadratic equations; seven eights did not get so far. In physics, the book used in recitations was Enfield's *Elements of Natural Philosophy*,—an insufficient and unskillful compend [sic] originally, and already far behind the progress of discovery. With the exception of Dr. Bowditch's contributions, the *Memoirs* of the American Academy [of Arts and Sciences] at that period betoken the same low state of science. There are papers not without value, but their value consists for the most part in judicious observations, or ingenious inventions and conjectures. The higher walks of mathematics and physical philosophy were untrodden.[91]

Even well into the nineteenth century, mathematical research in the United States was minimal. One reason was the difficulty of the

---

[91] John G. Palfrey, "John Farrar," *The Christian Examiner and Religious Miscellany*, 1853, *50*:121–136, on pp. 126–127.

subject, especially when compared to the relatively non-technical subjects in natural history. Many of these sciences required a keen eye and interest in the subject, but little formal training, as may be seen in this commentary from a would-be American geologist:

> I make no pretensions to any peculiar qualifications, other than that bodily health and constitutional fitness for labor and fatigue, which such an employment requires.[92]

Although the author probably goes too far in minimizing the requirements for geological study, the technical training was far less than that needed for research in mathematics and the physical sciences.

Yet, mathematics was even at a disadvantage with respect to some physical sciences such as astronomy and chemistry. Mathematicians did not enjoy the advantage of place which gave American astronomers incentive at least to make astronomical observations. The transit of Venus was important not because American astronomers were the only scientists with the capability of making such observations, but rather because Americans were in the right place to make them.

Chemistry achieved some degree of popularity in America in the late eighteenth and early nineteenth centuries for several reasons. Unlike mathematics, chemistry was presumed to be useful due to its applications in agriculture and manufacturing.[93] Chemistry was also tied closely to medicine, the most popular avocation of scientific-minded men in America. Many Americans traveling

---

[92] Amos Eaton, *An Index to the Geology of the Northern States, with Transverse Sections, Extending from the Susquehanna River to the Atlantic, Crossing the Catskill Mountains, to which is Prefixed a Geological Grammar*, ix. Quoted in Greene, *American Science*, p. 24.

[93] In *Science and the Ante-Bellum American College* (p. 96), Guralnick maintains:
> Until the widespread applications of steam power and electricity supplied new attention-getting alternatives, the social betterment derived from applications of chemistry to the industrial arts defined the claims of American science upon public support.

to Edinburgh to study medicine studied chemistry under Joseph Black while at the University.[94] The prominent colonial physician, Benjamin Rush, was professor of chemistry at the College of Philadelphia before becoming professor of medicine. In addition, Joseph Priestly's arrival in America in 1794 stimulated interest in chemistry.[95] Finally, the new theories that were revolutionizing the study of chemistry were more readily accessible to the public than were theories in mathematics.[96]

Despite the bleak outlook for American mathematics, several factors combined to allow a small group of dedicated men to lay the foundation for future generations. Leaders such as Robert Adrain, John Farrar, and especially Nathaniel Bowditch emerged. These men, and others like them, made the first attempts to organize and sustain journals dedicated to mathematics. Finally, they also adopted new methods in mathematics from the French during the first half of the nineteenth century, important for research and education of mathematics students. Each of these factors contributed to the development of what would become a community of professionals later in the century.[97]

---

[94] Reid-Maroney, *Philadelphia's Enlightenment*, p. 122.

[95] For a discussion of Priestly's influence on chemistry in America, see Greene, *American Science*.

[96] *Ibid.*, p. 172.

[97] See Parshall and Rowe, *Emergence*.

Chapter 3

A Prosopographical Analysis of the Early American Mathematics Publication Community[1]

To understand the context within which important foundations for the future growth of American mathematics were laid, it is important first to describe the community of mathematical practitioners that existed in America in the early nineteenth century. As was discussed in Chapter 1, the use of the term "community" to characterize the small, loose-knit group of Americans who engaged in various forms of mathematics is problematic. To alleviate this concern, this chapter will concentrate on a specific type of well-defined community—that of a publication community.[2]

---

[1] Material from this chapter has appeared previously in Todd Timmons, "A Prosopographical Analysis of the Earl American Mathematics Community," *Historia Mathematica*, 2004, *31*(Issue 4): 429–454.

[2] On the notion of a publication community as an analytical tool, see Derek de Solla Price, "Toward a Model for Science Indicators," in Yehuda Elkana, et. al., eds, *Toward a Metric of Science: The Advent of Science Indicators* (New York: John Wiley & Sons, 1978), pp. 69–96. For a study of a specific publication community, see Sloan Evans Despeaux, *The Development of a Publication Community: Nineteenth-Century Mathematics in British Scientific Journals*, Ph.D. dissertation, University of Virginia, 2002.

Of course, publication is only one metric by which a community may be measured. It is, however, an extremely important indication of the level of activity in a community of like-minded people. The beginning of a publication community in mathematics was an essential component for the future development of other aspects of American mathematics, most notably a community of mathematical researchers. Derek de Solla Price argues:

> The act of creation in scientific research is incomplete without publication, for it is publication that provides the corrective process, the evaluation, and perhaps the assent of the relevant scientific community.[3]

Although an American mathematical research community was the better part of a century away, the nascent publication community formed one of the foundations upon which this research community was built.

A study of the mathematics publication community in the United States in the early nineteenth century is a key to an analysis of the four prerequisite requirements for the development of a community of researchers discussed in Chapter 1. A numerical analysis of authors of mathematics[4] addresses the question of critical mass, while a study of their professions helps to answer the questions concerning financial support of the community. Within this study of the professions of the authors, a better understanding may be had of the amount of mathematical work occurring at the colleges. Finally, the entire notion of a publication community addresses the ability of the authors to exchange their mathematical ideas with people of similar interests.

---

[3] Price, "Toward a Model," p. 80.

[4] As discussed in Chapter 1, "mathematics" is used in its nineteenth-century context. Hence, astronomy, surveying, mechanics and other forms of applied, or mixed, mathematics are included in this survey.

One of the characteristics that marked the mathematics publication community in the period 1800–1838 was the reliance upon general science periodicals as an avenue for publication. Several short-lived mathematics journals appeared on the scene during this period, but their influence was minimal.[5] Three general science periodicals played an important role in providing an outlet for mathematical publications: the *Transactions of the American Philosophical Society* (henceforth referred to as the *Transactions*),the *Memoirs of the American Academy of Arts and Sciences* (the *Memoirs*),and *The American Journal of Science and Arts* (the *Journal*).

The first volume of the *Transactions* was published by the American Philosophical Society of Philadelphia in 1771. The Society, founded loosely on the original plan of Benjamin Franklin, became the first important scientific society in the United States. The *Transactions* appeared sporadically[6] and included articles and communications addressing a wide variety of sciences and liberal arts.

In 1785, the first *Memoirs* of the American Academy of Arts and Sciences appeared. The Academy, founded in Boston with John Adams as one of its organizers, also published a wide range of articles on scientific and other subjects. Like the *Transactions*, the *Memoirs* did not appear on a regular basis.[7]

The third publication included in this study, the *Journal*, was founded by Yale professor Benjamin Silliman in 1818. The most

---

[5] These mathematics journals, examined in Chapter 5, include *The Mathematical Correspondent* (1804–1807), *The Analyst, or Mathematical Museum* (1808), *The Analyst* (1814), *The Mathematical Diary* (1825–1832) and *The Mathematical Miscellany* (1836–1839). While the present chapter concentrates on the people who published mathematics in early America, Chapter 5 addresses in more detail the science and mathematics publications themselves.

[6] In the period 1771–1834, a total of ten volumes appeared. See Appendix 1 for the dates and mathematical content of each volume.

[7] From 1785–1833 a total of 8 volumes of the *Memoirs* were published by the Academy. See Appendix 1.

important differences between the *Journal* and the other publications mentioned here were its independence from a regional learned society and Silliman's determination to publish on a regular basis.[8] Each of these three journals is discussed in more detail in Chapter 4.

Although the period of American mathematics addressed in this study is 1800–1838, the data for this chapter includes the mathematics published in the United States from 1771–1834. I have chosen to begin in 1771 so that the full influence of the *Transactions* and the *Memoirs*, begun in 1771 and 1785, respectively, may be studied. The ending date of 1834 was chosen for two reasons. First, issues of the *Transactions* and the *Memoirs* nearly coincide (1833 and 1834, respectively) and secondly, 1834 marks the beginning of a drastic decline in the mathematical content of the *Journal*.[9]

3.1 A Statistical Survey of the American Mathematics Publication Community

The men who published mathematics in these three journals were a diverse lot geographically, professionally, and in terms of education.[10] In this diversity, however, we find common ground. Geographically, the majority of the publication community was clustered around a few population centers. Of the 81 American authors who published mathematics in the *Transactions*, the *Memoirs*, and

---

[8] Twenty-six volumes of the *Journal* appeared from its inception in 1818 through 1834.

[9] For further discussion of this decline, see Chapter 5.

[10] Several appendices contain data upon which this analysis is based. Appendix 1 contains a summary of mathematics by journal, total pages and type of mathematics (mixed, pure, etc.). Appendix 2 groups the authors according to the journal in which they published. Appendix 3 is a biographical summary of every American author who published mathematics in the *Transactions*, the *Memoirs* and the *Journal* between 1771 and 1834.

the *Journal* between 1771 and 1834, I have found information giving the residence of 64. Of the total, 19 resided in Massachusetts, 10 in Pennsylvania and 12 in New York. Therefore, approximately two-thirds of the authors were from only three states. This is certainly to be expected, as the population and intellectual centers of the nation were located in the northeast in general and in these three states in particular. It is also a consideration that all three journals were published in the northeast, the *Transactions* in Pennsylvania, the *Memoirs* in Massachusetts, and the *Journal* in Connecticut.[11]

Only a few authors came from outside the northeastern United States. For instance, one each hailed from Virginia, Tennessee, South Carolina, Ohio, and Louisiana. This paucity of participation from the south and west in the publication of mathematics is attributable to several factors. Population played a large role. Except for Virginia, which maintained a relatively large population and was adjacent to, if not in the middle of, the population centers of the country, the other states and territories outside of the northeast portion of the country had much smaller populations.

Another reason for the dominance of the northeast in mathematical publications is the traditional agrarian lifestyle of the other parts of the country. Education was not emphasized as it was in many parts of the rapidly industrializing north and mathematics suffered, as did other areas of learning and erudition. Whatever the reasons, it is obvious that the northeastern section of the young country dominated the mathematical life of the nation as it did its general intellectual life.[12]

---

[11] In Chapter 5 I discuss the regionalization of the individual journals.

[12] For discussion of the concentration of intellectual pursuits in specific areas of the United States, see Russel Blaine Nye, *The Cultural Life of the New Nation* (New York: Harper and Row, 1960). For the geographic distribution of scientific pursuits, see especially John C. Greene, *American Science in the Age of Jefferson* (Ames: The Iowa State University Press, 1984).

The education of our community of practitioners is also a telling statistic. Of the 48 on whom I have information concerning their educational background, 32 (two-thirds) graduated from college. Harvard graduates lead the way with a total of 11, followed by Yale with 8. No other college is represented more than two times.

However, one must be careful concerning the conclusions drawn from the data. The 33 on whom no education level is known could greatly change the analysis. For instance, if none of these 33 graduated from college, then the proportion of college graduates drops to slightly higher than one-third (39%) of the total population. On the other hand, if *all* of the 33 were college graduates, then the percentage rises to 80%.

Both the low number of 39% and the high end of 80% seem unreasonable. It is likely that at least a few of the men for whom no data is available were college graduates, but certainly not all of them. If it is assumed that a college degree increased the possibility of the type of success that would lead to inclusion in modern biographical works, then logic would dictate that few of these unknown contributors possessed such a degree. With this assumption, a number between 40% and 50% might form a working conjecture for the proportion of college graduates from the population of authors.

It must be remembered, though, that college graduation did not necessarily correlate to mathematical abilities. Long before the elective system appeared in the United States, all college graduates studied a smattering of mathematics. Those who desired to pursue further studies in mathematics were limited to their own devices with possibly a small amount of guidance from an interested professor.

The concentration of the mathematics authors in the northeastern section of the country might appear to prepare a fertile ground for the development of a community of mathematicians. Several factors militate against this concentration, however. Perhaps the most important is the lack of professionalization in the community of mathematics authors.

Although thirty-three of the authors (41%) listed were at one time or another professors at an American college, this statistic may be misleading. First of all, the tenure of many of the authors was a short one. George Baron, for instance, was professor of mathematics at West Point for only one year. David Rittenhouse's tenure at the University of Pennsylvania lasted only three years. In fact, of the 28 professors whose length of service is known, the employment of 10 of the group lasted five years or less. Long careers as professors of mathematics (or science) were not common in early America.

Another reason to believe that the large number of professors is not as significant as it might seem is related to the first. Many of the authors were employed in careers outside of academia before, during, and after their tenure at their respective colleges. In addition to their positions in education, several were engineers or surveyors (Edwin Johnson, Andrew Ellicott, Ferdinand Hassler, and Alden Partridge). There were college administrators (Samuel Webber and Robert Patterson), an actuary (Elizer Wright), an almanac-maker (Benjamin West), a clergyman (William Smith), and a physician (Hugh Williamson). There was also an inventor, a blacksmith, a naval officer, and several merchants. All in all, this eclectic group did not represent the modern idea of a professional community of scholars.

When considering the relative importance of mathematics in early America and the time available to these professors for mathematical work, it is also important to realize that before mid-century it was very rare for a college to have a professorship dedicated strictly to mathematics. The vast majority of the 33 professors who published mathematics in one of the three major journals were

Figure 3.1.1: Theodore Strong

professors of mathematics and natural philosophy, or some other combination involving mathematics and the sciences. Of the 33, only 10 were listed as professors of mathematics exclusively; and 6 of the 10 represent one school, the Military Academy at West Point.[13]

This dual role had a two-fold affect on the quantity and quality of mathematics appearing in the journals. First of all, a professor of mathematics and natural philosophy had duties in mathematics as well as astronomy, chemistry, physics, and possibly other sciences. Obviously, this left little time for concentrated efforts in any one area. Secondly, some of these men were professors of mathematics in name only, having interests that actually lay outside the area of mathematics.[14]

If 33 of the 81 authors in this study were professors at American colleges, that leaves over half of the total in professions other than college professor. In addition to the men mentioned above who split their time between professorships and other forms of employment, many of the authors spent their whole careers as surveyors, engineers, actuaries, businessmen, clergymen, teachers at Latin and other schools, instrument-makers, college administrators, lawyers, and physicians.

Eighty-one men publishing mathematics in three scientific journals seems, at first glance, to be a significant number. There are, however, certain mitigating factors to consider. The time period in question begins in 1771 with the first volume of the *Transactions* and ends in 1834. A total of 63 years makes the 81 total mathematical authors seem less significant.

---

[13] For a discussion of the high turnover rate of professors at the young military academy, see Joe Albree, David C. Arney and V. Fredrick Rickey, *A Station Favorable to the Pursuit of Science* (Providence, RI: American Mathematical Society, 2000).

[14] See discussion below concerning Parker Cleaveland.

Of course, the *Transactions* and the *Memoirs* were not published regularly throughout those 63 years, and the *Journal* did not begin publication until near the end of the period. Even so, a total of 44 volumes[15] of the three journals appeared in the 63 years. Eighty-one authors spread over 44 volumes remains a relatively small population.

As discussed previously, the majority of the 81 authors were not involved to any great degree with mathematics or mathematics education. In fact, only nine of the 81 might be considered as having had any kind of significant influence on mathematics in the United States for the seven decades of this time period.[16] These major figures were significant for their publication record, their teaching career or for other contributions to American mathematical sciences.

3.2 Major Figures in Early American Mathematics

One of these men, David Rittenhouse, was discussed at length in the previous chapter. Three others, Nathaniel Bowditch, Robert Adrain, and John Farrar are studied in detail in later chapters. A fourth man, Charles Davies, more appropriately belongs to the next generation of American mathematicians. Some of Davies' early publications appeared in *The American Journal of Science and Arts* near the end of our period, but his most important work came later as the author of a series of influential textbooks.[17] The

---

[15]This includes 10 volumes of the *Transactions*, eight volumes of the *Memoirs*, and 26 volumes (in 17 years) of the *Journal*.

[16]In his study of publication patterns in the scientific community, de Solla Price, "Toward a Model," finds that in any publication community, a relatively small core of authors publish consistently over a period of time. The rest of the publications in a given community come from what he calls transient authors, those who publish sporadically at best. This is certainly an accurate portrayal of the nascent mathematics publication community in the United States.

[17]See Chapter 6 for further discussion of Davies and his textbooks.

additional four men warrant further discussion for their influence on American mathematics.

Theodore Strong (1790–1869) was a professor of mathematics at Hamilton College (New York) from 1812–1827, then at Rutgers from 1827–1863.[18] He formed an important link in the chain of American mathematics, as his teacher was the Yale mathematics professor Jeremiah Day and one of his students was George William Hill, a noted American mathematician and astronomer. Strong published mathematics in several journals, including the *Mathematical Diary* and the *Mathematical Miscellany*, two short-lived mathematical journals discussed further in Chapter 5.

Strong was one of the most prolific contributors to the new *Journal*, founded by Benjamin Silliman in 1818. During the period 1818–1834, Strong published 20 articles in Silliman's *Journal*.[19] Included among these was the very first mathematical article published in the journal.[20]

It is sometimes difficult to discern the motivation behind many of the mathematical papers appearing in the early American journals. Some appear to have as their primary purpose the solution of mathematical problems for the benefit of students of mathematics. One such paper appeared in the *Journal* in 1820 under the authorship of Strong. In this work,[21] Strong laid out eight problems in

---

[18] Biographical information on Strong may be found in Edward R. Hogan, "Theodore Strong and Ante-Bellum American Mathematics," *Historia Mathematica*, 1981, *8*:439–455.

[19] Further analysis of Strong's contributions to this journal may also be found in Chapter 5.

[20] Theodore Strong, "An improved Method of obtaining the Formulae for the Sines and Cosines of The Sum and Difference of two Arcs," *American Journal*, 1819, *1*(No. 4):424–427. Strong presents his own unique method for deriving the familiar formulas for $\sin(A \pm B)$ and $\cos(A \pm B)$.

[21] Strong, "Mathematical Problems, with geometrical constructions and demonstrations," *American Journal*, 1820, *2*(No. 1): 54–64.

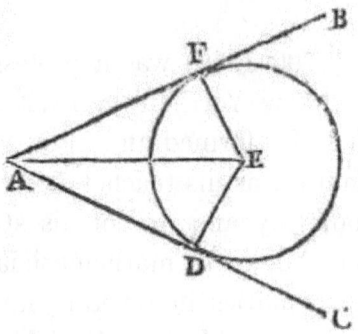

Figure 3.2.1: Geometry Problem

Euclidean geometry, the solutions of which would have been easily followed by any student with a smattering of training in geometry. For instance, in problem IV, Strong stated:

> Let two straight lines and a point which does not lie at the intersection of those lines, be given in position, it is required to describe a circle through the given point to touch the two given straight lines.[22]

For case I, in which the given point lies on one of the given lines, Strong supplies Figure 3.2.1.

Line AE bisects the angle BAC, and lines DE and FE are perpendicular to AC and AB, respectively. Then a circle with radius DE is the required result. Strong closed by giving a brief proof of the veracity of his construction. This is certainly not an important result in Euclidean geometry, but rather indicates an intention to educate the reader concerning the basics of Euclidean construction and proof. This fits the pattern of many of the articles authored by Strong.

---
[22] *Ibid.* p. 56.

Another important figure in the early American scientific community was the Swiss immigrant, Ferdinand Hassler (1770–1843) who came to the United States in 1805.[23] Already having established his reputation as a first-rate surveyor, Hassler was recommended to head the team to survey the United States coast. Several factors conspired to delay the establishment of the United States Coast Survey until 1816, at which time Hassler was named superintendent.[24] In the meantime, Hassler served brief stints as professor of mathematics at West Point and as professor of mathematics and natural philosophy at Union College. Although Hassler published only two pieces in the *Transactions*, one of these articles was a large collection of his papers concerning his work with the Coast Survey (1825).

Hassler was only one of many immigrants who contributed greatly to American mathematics. Robert Patterson (1743–1824) was born in Ireland and immigrated to the United States in 1768.[25] After spending several years in business and in teaching navigation in Philadelphia, Patterson was appointed professor of mathematics at the University of Pennsylvania in 1779. In 1814 Patterson became vice-provost of the university and his son, Robert M. Patterson, succeeded him as professor of mathematics.

Patterson was an important contributor to the *Transactions*, authoring a total of 11 articles. These articles ranged from astronomical observations to descriptions of mechanical devices. In the

---

[23] For biographical information, see Florian Cajori, *The Chequered Career of Ferdinand Rudolph Hassler, first superintendent of the United States Coast Survey* (Boston: The Christopher publishing house, 1929).

[24] Politics intervened, and Hassler was to spend most of the rest of his life in and out of the employment of the U.S. Coast Survey. See Cajori, *Chequered* and A. Hunter Dupree, *Science in the Federal Government: A History of Policies and Activities to 1940* (Cambridge, MA: Belknap Press, 1957).

[25] Biographical information comes primarily from *Dictionary of American Biography*, Volume 14, pp. 305–306.

spirit of the medieval quadrivium, Patterson even contributed an article concerning musical notation.

Andrew Ellicott (1754–1820) was a student of Robert Patterson when Patterson taught navigation and surveying in Philadelphia from 1768 to 1772.[26] Ellicott became an accomplished surveyor, spending the first years of his professional career in various government-appointed posts surveying the new land, including the newly established District of Columbia. From 1813 until his death, Ellicott served as professor of mathematics at West Point. Ellicott contributed a total of 15 articles to the *Transactions*. Each of these articles concerned his work in surveying or his astronomical observations and calculations.

In addition to the nine mentioned above, several other men who published mathematical works prior to 1838 deserve mention by name. Two, Samuel Williams and Samuel Webber, although not important for any significant contributions to mathematics, did hold the most prestigious mathematics position in the United States, the Hollis Chair of Mathematics and Natural Philosophy at Harvard. Webber eventually became president of Harvard.

Alexander M. Fisher, professor of mathematics and natural philosophy at Yale from 1817–1822, is an interesting, but tragic, character in American science. Fisher was killed in a shipwreck in 1822. His obituary extols Fisher as an outstanding teacher and a curious scientist;[27] Robert Adrain referred to him as "the very ingenious and much lamented Professor Fisher."[28] When he was killed, Fisher was enroute to visit the great scientists of France, in the hopes of learning their new techniques and bringing them back to America. His contemporaries knew Fisher as a man "With a

---

[26] Biographical information comes primarily from *Dictionary*, Volume 6, pp. 89–90.

[27] "Obituary of Professor Fisher," *Journal*, 1822, 5:367–376.

[28] *Mathematical Diary*, 1825, *1*:iv.

mind so unshackled [to traditional and ancient dogma], he was in a high degree prepared for original investigation."[29] His impact on American mathematics, had he lived, cannot be known. However, his abilities and interests certainly had Fisher pointed in the direction of leading this generation of American mathematicians into the new age of modern mathematics.

Two men who perhaps best epitomize the "mathematician" of the early nineteenth century were Parker Cleaveland and A.B. Quinby. Cleaveland (1780–1858) was professor of mathematics and natural philosophy at Bowdoin College (Maine) from 1805 until his death.[30] In spite of his title, Cleaveland was not really a mathematician, or for that matter a physical scientist. He was a noted mineralogist who published an influential textbook entitled *An Elementary Treatise on Mineralogy and Geology* (1816). Although Cleaveland contributed several articles to the *Memoirs* and to the *Journal* on subjects in mixed mathematics such as astronomical observations and mechanics, his influence was certainly not great as a member of the mathematical publication community.

Cleaveland was one of the few men in the population of American mathematics authors who published work in more than one of the general science journals.[31] None published work in all three journals. Interestingly, all but two of these men included the *Journal* as one of their publication sources. The two who did not, Benjamin West and Samuel Williams, were both dead before the *Journal* was founded. This indicates that the regularity of publi-

---

[29] *Journal*, 1822, 5:372.

[30] Biographical information comes primarily from *Dictionary*, volume 4, pp. 189–190.

[31] The complete list, with the journals in which they published, is as follows: Nathaniel Bowditch (*Memoirs* and *Journal*), Parker Cleaveland (*Memoirs* and *Journal*), James Dean (*Memoirs* and *Journal*), Alexander Fisher (*Memoirs* and *Journal*), Benjamin West (*Transactions* and *Memoirs*), Samuel Williams (*Memoirs* and *Transactions*) and Elizur Wright (*Memoirs* and *Journal*).

cation of the *Journal* played an important role in the publication community.

Another notable statistic is that, excluding West and Williams, the remaining five men all published in the *Memoirs* and the *Journal*. This suggests a continuing regionalization of the journals, in spite of Benjamin Silliman's attempts to make the *Journal* a national publication. The proximity of the Massachusetts-based *Memoirs* and the Connecticut-based *Journal* led to a natural intermingling of their respective authors.

In addition to Parker Cleaveland, another noteworthy author who does not fit the modern definition of "mathematician" was A.B. Quinby, one of the most mysterious men of this publication community. Little is known about Quinby. He contributed numerous articles to the *Journal*, but does not seem to have published anywhere else. His interests lay chiefly in application of mathematics to the mechanical arts, and his contributions were usually of a comparatively sophisticated mathematical nature.

For a brief time, Quinby was one of the most prolific contributors to the pages of the *Journal*. His first article appeared in 1824 and was entitled "On Crank Motion."[32] In this article, Quinby addressed a claim by a previous contributor to the *Journal*:

> Mr. Ward says, "The pressure of the steam upon the piston being uniform through the stroke; it follows that the impulses (I understand upon the upper end of the shackle-bar) at all times are equal to one another; and this being the case, it is equally a matter of course that the effects produced at the several points of division of the quadrant, are as the perpendiculars respectively from those points to the line of force."[33]

---

[32] *Journal*, 1824, 7:316–322.

[33] *Ibid.*, p. 316.

Quinby constructed, using arguments from mechanics and from Euclid, a proof by contradiction that

> the effects ... at the several points of division of the quadrant are *not* to one another, as the perpendiculars respectively from those points to the line of force.[34]

Quinby went on to construct his own mechanical proof that *"the crank motion occasions no loss of the acting power."*[35] Throughout his demonstrations, Quinby exhibited a thorough knowledge of Euclidean geometry and mechanics, traits remarkable in a man who has left so little evidence of his very existence.

Between 1824 and 1828 A.B. Quinby contributed 12 studies to the *Journal*, all of a mechanical nature. In addition, seven other articles addressing (and sometimes refuting) Quinby's work were published. Although his identity continues to be a mystery, his presence in the pages of the *Journal* during the mid-1820s is notable.

As exemplified by Quinby's contributions, the application of mathematics to mechanical processes was an underlying current throughout the pages of the *Journal*. In addition to Quinby's work, articles concerning mechanical processes were written by such men as Eli W. Blake, a Connecticut firearms manufacturer; a physician by the name of Felix Paucalis; the aforementioned Parker Cleaveland and others. A look at a few of these publications, falling under the rubric of mixed mathematics, will serve to demonstrate the types of problems that interested American mathematical practitioners.

---

[34] *Ibid.*, p. 318.

[35] *Ibid.*

3.3 Mixed Mathematics in American Journals

Some professions, such as surveyor, engineer or actuary, obviously lend themselves to mathematical ability. Surveying played an especially important role in mathematics. A new country with vast, unmapped territory required men proficient in the use of mathematical instruments and astronomical observations to establish its boundaries. Many of the men who published mathematical papers in America depended primarily upon surveying to earn a living. Andrew Ellicott, Ferdinand Hassler, and Nathaniel Bowditch all enjoyed reputations as careful and ingenious surveyors. Many of the articles that appeared in the *Transactions*, the *Memoirs*, and the *Journal* were either directly related to surveying, or to the astronomical observations required to establish the exact position of American locales.

Two examples of such papers serve to establish the type of communications that are categorized as "mixed" mathematics. In 1815 there appeared in the *Memoirs* an article from a Massachusetts resident by the name of Epaphras Hoyt.[36] Hoyt's work, communicated in a letter to Harvard professor John Farrar, listed various observations made by Hoyt with a 30-inch telescope, a micrometer, a "very good metal clock, with a second hand," a 10-inch sextant and "an accurate compass of the Rittenhouse construction," among other instruments.[37] From his observations, Hoyt calculated the longitude of the town of Deerfield from separate observations. These longitude calculations ranged from $72°21'30''$ found from an observation of a solar eclipse to $72°26'15''$ found "from the distance of the Moon and Aldebaran, observed with the sextant on Oct. 29,

---

[36]"Astronomical observations made near the center of the village of Deerfield, Massachusetts," *Memoirs*, 1815, *3*(Part 2):305–307.

[37]*Ibid.*, pp. 305–306.

1811."[38] Hoyt's communication included neither sample calculations nor mention of mathematical techniques.

A second example of the publication of such raw astronomical data came from the pen of the well-known surveyor, Andrew Ellicott.[39] Like Hoyt's paper three years before, the bulk of Ellicott's article is simply observations of such occurrences as an eclipse of the moon, eclipses of Jupiter's satellites, and a comet that appeared in 1807. Ellicott closes his communication with a formula "for calculating the parallax in latitude, and longitude" that he attributes to the British *Astronomer Royale*, Dr. Maskelyne.[40]

Not all of the astronomical papers appearing in the pages of American periodicals were simply compilations of raw data. Nathaniel Bowditch, for one, published numerous astronomical papers of a mathematically sophisticated nature.[41] Another example of a more advanced theoretical study came from James Dean, professor of mathematics and natural philosophy at the University of Vermont. In an article appearing in the *Memoirs* in 1815, Dean argued that,

> The inequality of the moon's motion about the earth, combined with the effect of the inclination of the lunar orbit and equator (which cause the moon's librations) produce to a spectator placed on the surface of the moon, an apparent motion of the earth about its mean place, supposed at rest.[42]

---

[38] *Ibid.*

[39] "Astronomical Observations, &c.," *Transactions*, 1818, *1*(new series):93–102.

[40] *Ibid.*, pp. 98–99.

[41] See Chapter 4.

[42] "Of the apparent motion of the earth viewed from the moon, arising from the moon's librations," *Memoirs*, 1815, *3*(Part 2):241–245, on p. 241.

Dean based his geometric argument on the inclination of the lunar orbit and several carefully stated propositions related to the relative location of the moon in its orbit of the earth.

3.4 Pure Mathematics in American Journals

In Chapter 2, the importance of utility to early American science was discussed at length. Mathematics in the early nineteenth century was likewise valued primarily for its usefulness. Although pure mathematics received little attention in the first century of America's existence, a few authors did attempt to communicate their interests in the field. One such contributor, whose identity remains a mystery, is one C. Wilder, identified only as "of New Orleans." In 1829 Wilder published a rather sophisticated article (at least for the time and place) in the *Journal* entitled simply "Algebraic Solutions."[43] In this article, the author demonstrated a general method for finding solutions of equations of the form

$$y^n + ay^{n-1} + by^{n-2} + cy^{n-3} + dy^{n-4} \cdots + ky + l = 0$$

by utilizing a rational function of the form

$$\frac{x^{m(n-1)} + S_m x^{m(n-2)} + S_{2m} x^{m(n-3)} + S_{3m} x^{m(n-4)} + \cdots + S_{(n-2)m} x^m + S_{(n-1)m}}{x^{n-1} + yx^{n-2} + px^{n-3} + qx^{n-4} + \cdots + tx + u}$$

such that the denominator of the function is a factor of the numerator.

For example, to solve $y^2 + ay + b = 0$, Wilder began with the rational function $\frac{x^2 + S_2}{x+y}$ and satisfied the requirement that the denominator be a factor of the numerator by letting $S_2 = -y^2$. By making this substitution and another, $y = y + z$, the rational function becomes

$$\frac{x^2 - y^2 - 2zy - z^2}{x + y + z}.$$

---

[43] *American Journal*, 1829, *16*(No. 2):271–282.

Wilder then noted that the numerator, as a function of y, has the form
$$y^2 + ay + b.$$
By equating $x^2 - y^2 - 2xy - z^2$ to $y^2 + ay + b$, Wilder noted that $2z = a$ and $z^2 - x^2 = b$. Solving these last two equations along with $x + y + z = 0$, Wilder found that
$$y = \frac{-a}{2} - x,$$
where
$$x^2 = \frac{a^2 - 4b}{4},$$
which the author calls "the common rule [for a solution of the quadratic equation]".

Wilder continued his demonstration by solving several more higher-order equations using the same method. For instance, to solve $y^4 + by^2 + cy + d = 0$, he employed the rational expression
$$\frac{x^{12} + S_4 x^8 + S_8 x^4 + S_{12}}{x^3 + yx^2 + px + q}.$$
After providing several more examples, Wilder then proceeded to lay out a general method for finding solutions to any equation of the form given.

Wilder referred to the methods given in Lacroix's *Algebra*. He may have consulted Lacroix's book in French, or he may have studied the English translation of Lacroix made by Farrar at Harvard over a decade earlier. Whatever the case, Wilder's contribution to the *Journal* represents one of the most interesting mathematical works in the pages of an American publication of the period.

Another article in the field of algebra appeared in the *Journal* in 1831, contributed by A.D. Wheeler, the principal of the Latin Grammar School in Salem, Massachusetts. One thing that makes this article interesting is that it is one of many appearances in

American journals of a problem in Diophantine Algebra, a subject popularized by Robert Adrain in several short-lived mathematical journals founded by Adrain himself.[44]

Wheeler stated the problem:

> To find two squares, whose sum shall be square; or in other words, to find rational values for the legs and hypothenuse [sic] of a right angled triangle.[45]

Wheeler proceeded to give a rule for finding such numbers that answer his proposed problem:

> Take any two numbers, of which the difference is 2. Their *sum* will be the *root* of one square; their *product*, that of the other. Add 2 to the product, just found, and you obtain the root of the *sum of the squares*, or the value of the hypothenuse [sic].[46]

The author gave the example of 10 and 12. The sum of the square of their sum and the square of their product is equal to the square of their product plus two:

$$10 + 12 = 22,$$
$$0 \times 12 = 120, \text{ and}$$
$$22^2 + 120^2 = (120 + 2)^2.$$

Wheeler followed with a straightforward algebraic proof of his rule, and then supplied several additional examples, noting other interesting numerical properties of the numbers thus obtained. This is certainly not an earth-shattering result, but it is interesting as

---

[44] See Chapter 5.

[45] "An Easy Solution of a Diophantine Problem," *American Journal*, 1831, *20*(No. 2):295–297.

[46] *Ibid.*, p. 295.

an example of pure mathematics in a time when American science and mathematics continued to emphasize the useful aspects of their findings.

Unfortunately, two forays into the field of pure mathematics were embarrassing attempts to solve ancient—and by this time, dying—problems. James Winthrop published the "solutions" to the ancient problems of Greek geometry in the *Memoirs* in 1793. Winthrop (1752–1821) was the son of the well-known Harvard professor, John Winthrop.[47] James Winthrop was more than once considered for his father's position at Harvard, but each time was rejected due in part to his "intemperate manner and his eccentricities."[48] Winthrop was at various times librarian at Harvard, registrar of probate for Middlesex, and postmaster of Cambridge.

Winthrop's first ill-advised article, "Geometrical Methods of finding any required Series of Mean Proportionals between given Extremes,"[49] was an attempt to solve the problem of duplicating the cube. Although Winthrop's mathematics is not of interest, the fact his attempt to solve the problem appeared in an American journal is significant. His work instigated a public encounter with George Baron.[50] Baron responded to Winthrop's article in the next publication of the *Memoirs*, the very year that Baron began the first American periodical dedicated to mathematics.[51] In his response, Baron pointed out the fallacy in Winthrop's argument.

Interestingly, in a footnote to Baron's paper, the editor of the *Memoirs* admitted that upon examination of Winthrop's paper on

---

[47] Biographical information comes primarily from *Dictionary*, volume 20, pp. 407–408.

[48] *Ibid*, p. 407.

[49] *Memoirs*, 1793, 2:9–13.

[50] Baron, a recent English emigrant, taught briefly at West Point and founded his own academy in New York.

[51] See Chapter 5 for further discussion of this and other mathematical journals.

arrival at the Academy, several members "skilled in mathematical science" thought it false. "But at the particular request of the author the committee for publication consented to its insertion in the *Memoirs*."[52] A refereed journal, this was not!

Winthrop's other contribution to the same volume of the *Memoirs* was another attempt at solving an ancient problem, that of trisecting an angle.[53] Once again Winthrop's argument failed to hold true, although this work did not seem to draw the ire of others, as had his previous debacle.[54]

This discussion has by no means included all of the 81 men who published mathematical articles in the *Memoirs*, the *Transactions* and the *Journal*. Other notable names appear in the pages as authors. These include William Cranch Bond, the first director of the Harvard Observatory. Bond, trained as a watchmaker, studied astronomy in his spare time. His association with Harvard began in 1815 when the college sent him to Europe with the hopes of gathering information pertinent to a planned Harvard observatory. However, the observatory would not come to fruition until 1839. As director, Bond made several discoveries that established his international reputation. He was a pioneer in the use of photography in astronomical observations, and in partnership with his son, George Phillips Bond, discovered the eighth moon of Saturn and the only moon of Neptune.

Also in the list of authors is Jared Mansfield, whose book, *Essays, Mathematical and Physical*[55] was one of the first science

---

[52] *Memoirs*, 1804, 2:42.

[53] "A Rule for Trisecting Angles geometrically," *Memoirs*, 1793, 2:14–17.

[54] For discussions of the history of these ancient Greek problems, including other futile attempts at solutions and their eventual resolution, see Douglass M. Jesseph, *Squaring the Circle* (Chicago: University of Chicago Press, 1999); Petr Beckmann, *A History of Pi* (New York: Barnes & Noble Books, 1971); and Thomas Heath, *A History of Greek Mathematics* (New York: Dover Publications, Inc., 1981).

[55] New Haven, CT: William W. Morse, 1802.

textbooks published in the United States. Mansfield worked as a teacher and a surveyor, and served in the U.S. Army before becoming professor of natural philosophy at West Point in 1812, where he stayed until 1828.

Authors also include Joseph Willard, president of Harvard, and two other professors of natural philosophy at American colleges who published a significant quantity of work in these journals: Elizur Wright at Western Reserve College in Ohio and Denison Olmsted at the University of North Carolina and later Yale.

Wright (1804–1885) was a professor of mathematics and natural philosophy at Western Reserve College in Ohio and later an actuary for several insurance companies; however, he was best known as a leader in the anti-slavery movement.

Olmsted (1791–1859), a graduate of Yale, spent most of his career in academic positions, first at the University of North Carolina and later at his alma mater. While in North Carolina, he performed the first state geological study in the United States. He moved to Yale in 1825 to accept the chair of mathematics and natural philosophy. During his career, Olmsted published a long list of popular textbooks on natural philosophy and astronomy.

The rest of the names appearing on the list of authors for the three journals are nondescript, to say the least. The level of mathematical sophistication and the fact that none were able to sustain any sort of consistent publication record indicates that none belonged to a community of *research* mathematicians. The number of men whose mathematics was of a higher level and who did sustain a significant period of work was small. Nonetheless, we have examined a *publication* community of authors with widespread geographic and mathematical bases in order to attempt to piece together characteristics of the group.

The typical member of this publication community was a man from Massachusetts, Pennsylvania or New York, with an education from Harvard or Yale. The author was most likely a college professor at one time in his career, although his tenure may have been short. He probably published in only one of the three important scientific journals, two at the most. If he achieved any sort of international reputation, it was for surveying, astronomy, or other form of mathematical application, not pure mathematics. Of course, this composite picture of the average mathematics author derives from a relatively small population and would undoubtedly have a rather large standard deviation. This small population is perhaps the most telling of the statistics concerning the American mathematical publication community.

Then the answer to the question, "Did America fail to form a sustained research community in mathematics in the early nineteenth century due to a lack of a critical mass of trained and interested mathematicians?" is an unequivocal yes. Eighty-one accomplished and dedicated mathematicians would seem short of the number needed. When only a half-dozen or so of that number are actually qualified (and marginally qualified at that) to carry out the duties of a research community, the possibility seems remote of just such a community developing.

As we have seen, the lack of a critical mass may be attributed to several factors. The lack of financial support in positions allowing time for mathematics is an overriding factor. The few men who did obtain positions as professors were, without exception, spread to thinly in their duties to have any real opportunity for developing any latent mathematical talents. As we saw in the case of Parker Cleaveland, the title "Professor of Mathematics and Natural Philosophy" did not even necessarily entail interests or abilities in mathematics.

This lack of concentrated mathematical talent in the colleges also made it unlikely that able students would receive the training or encouragement needed to develop research mathematicians. The work of Theodore Strong and John Farrar (see Chapter 6) began the process of training American mathematicians, but it would require another generation before their students would form the number of branches needed to sustain a viable tree of dedicated researchers.

Finally, although this chapter has dealt with a publication community contributing to three science journals, general interest periodicals (and in the case of the *Transactions* and the *Memoirs*, sporadically-published journals) were not enough to encourage or support a sustainable community. The first attempts to establish specialized mathematical journals were made in the first decades of the nineteenth century, but again it would require the work of another generation in the later part of the century to see the creation of permanent and high-quality mathematical journals.

In the meantime, American scientists sought leadership in much the same way the population hungered for political leadership. One man fulfilled this leadership role in the physical sciences. This man, Nathaniel Bowditch, serves as a case study for the community of prospective American mathematicians.

Chapter 4

An Emerging Leader: Nathaniel Bowditch and American Mathematics

Being a mathematician in America in the early nineteenth century was much different from being a mathematician in Europe during the same period. For that matter, any scientific endeavor in America was confronted with obstacles quite different from those faced in Europe. Opportunities that existed in the older societies of England, France, Germany, and other European nations did not exist in the young nation of America. European scientists and mathematicians had access to positions in royal courts and academies, as well as in the long-established colleges. These positions provided European scientists with employment opportunities in their chosen profession and, although not universally geared towards research,[1] often provided the one critical ingredient needed to produce original work: time.

---

[1] Many positions in European colleges were, like their American counterparts, teaching positions providing little time or incentive for research. Unlike the American colleges, however, there were distinguished chairs that provided comfortable employment with few teaching commitments, allowing the chair holder time for original research.

American mathematicians of the early nineteenth century did not have these same opportunities. Advanced training in mathematics was virtually non-existent, as the colleges in America had neither the teaching talent, nor the time, nor the motivation, economic or otherwise, to teach higher mathematics. Even if more favorable conditions had existed in the colleges, the American student was typically ill-prepared to embark upon such a course of study.

But what if an American student had the desire and the opportunity to study mathematics, or some other area of science? The prospects for supporting oneself while performing original research were bleak, if not altogether non-existent. A man of science in America in the early nineteenth century, unless independently wealthy, would have been forced to support himself in a peripheral field, such as medicine or teaching. Without the patronage of royalty, before the development of the research university, and long before the American university became the recipient of large amounts of outside funding targeted at research, an American mathematician was forced to pursue his research after the day's work was done.

Americans of this time period were only too aware of the conditions in their country that placed obstacles in the path of original research. They were also aware of the advantages held by European scientists. John Pickering, in a eulogy of Nathaniel Bowditch, compared Bowditch's lifetime quest for scientific discovery to that of Pierre de Laplace. Pickering noted that Laplace had

> ...the entire command of his time, and [was] surrounded by all the scientific men of France, who could render him any aid in their respective departments. If an observation in astronomy was required,—if any experiment became necessary in meteorology, in chemistry, in mechanics,—if laborious calculations were wanted in mathematics,—in

Figure 4.0.1: Nathaniel Bowditch

> order to verify his theories,—the most eminent men of France, at the most advanced period of human knowledge, may be truly said to have been at his command; some of them, indeed, literally so, by orders of the government; and others, from that common zeal in the cause of science, which is always glowing in such a community.
>
> And here, I cannot but ask you, for a moment, to compare with these highly favorable circumstances, the disadvantages, under which our lamented President [Bowditch] intrepidly undertook the difficult and laborious task [of translating Laplace's *Mécanique Céleste*], which he has so successfully accomplished.[2]

Pickering continued with a brief summary of the struggles Bowditch faced. He pointed out that, although born in humble circumstances, Laplace was discovered by D'Alembert thus setting into motion actions that resulted in Laplace's appointments to positions that gave him the leisure required to pursue his investigations. Bowditch, on the other hand, whose talents were also well known to influential men, never secured such a position and therefore was never "relieved from all solicitude for those necessary means of living."[3] The reason for Bowditch's inability to secure such a position was simple: no such position existed in the United States in the early nineteenth century.

Of course, science was not alone in its struggle to establish a place in the intellectual milieu of the developing country of America. The first volume of a new and ambitious literary journal lamented:

---

[2] John Pickering, "Mr. Pickering's Eulogy on Dr. Bowditch, President of the Academy," *Memoirs of the American Academy of Arts and Sciences*, 1846, Series 2, 2:xl. Bowditch was president of the American Academy of Arts and Sciences from 1829 until the time of his death in 1838.

[3] *Ibid.*, p. xlii.

Genius in composition, like genius in every other art, must be aided by culture, nourished by patronage, and supplied with leisure and materials. The genius of the poet, orator, and historian, cannot be exercised with vigour and effect, without suitable encouragement, any more than that of the artist and mechanic. Neither the one or the other is beyond the sphere of social affections, and domestic duties and wants; neither can be expected to produce works of ingenuity and labour without such a recompense as the natural ambition of man, and the necessities of his nature and situation demand.[4]

It was in this cultural setting that a young Nathaniel Bowditch began his life-long pursuit of science.

Nathaniel Bowditch (1773–1838) was the leading mathematician and astronomer in America during the first quarter of the nineteenth century. Although Bowditch was not in the same class as the leading scientists of Europe, he did build a reputation as an able mathematician, a talented astronomer, and an important leader in the nascent American scientific community. Bowditch published two major works, *The New American Practical Navigator* and a translation of and commentary on Laplace's monumental work, *Mécanique Céleste*. In addition, Bowditch published numerous papers in American journals and periodicals, particularly the *Memoirs of the American Academy of Arts and Sciences.*

Bowditch's reputation in America and abroad grew throughout his lifetime. With his abilities as a mathematics practitioner already confirmed with the publication of *The New American Practical Navigator* in 1802, Bowditch further enhanced his reputation among Europe's elite scientists with his translation of and commentary on Laplace's *Mécanique Céleste*, published between 1829–

---

[4] *American Review and Literary Journal*, 1801, *1*:iii–iv.

1839. Throughout his lifetime, Bowditch developed a correspondence with many scientists and mathematicians in America and in Europe. His first international correspondence seems to be a letter to the French mathematician, Sylvestre Lacroix, in which Bowditch noted some typographical errors in Lacroix's *Calculus*.[5] Nearly a year later Bowditch received a grateful response from Lacroix.[6]

Throughout his career, Bowditch was looked to for leadership from other American scientists. In a lengthy letter dated August 13, 1817,[7] Benjamin Vaughan[8] asked Bowditch to call to the attention of European writers errors in their works. Vaughan believed that in so doing Bowditch could earn greater respect for American science. Furthermore, Vaughan expressed his dismay that American scientists were so ignored even in America. He pointed out to Bowditch that only an incomplete collection of Franklin's scientific works had been published, and no publication of Count Rumford's works had been made available in America. He lamented the fact that "[David] Rittenhouse is neglected also." The content and tone of Vaughan's letter suggests that he believed that Bowditch, with his reputation in Europe well established, might help to remedy the unfortunate situation in which American science found itself.

---

[5] Letter dated September 18, 1805, in *Bowditch Collection*, Boston Public Library.

[6] Letter dated September 1, 1806, in *Ibid*.

[7] *Ibid*.

[8] Vaughan was an Englishman who migrated to America in 1797 due to political turmoil in England. Although not educated as a scientist himself, his interests in science led him to establish correspondence with many of the leading scientists of America and Europe.

4.1 Early Life

Nathaniel Bowditch was born in Salem, Massachusetts on March 26, 1773.[9] Bowditch spent his teenage years as an apprentice to a ship's chandler in Salem. During this time, Bowditch was constantly borrowing books from various Salem sources and educating himself in Latin (primarily so that he could read Newton's *Principia*), French,[10] mathematics, and the sciences. As luck would have it, Bowditch gained access to a fine collection of scientific works owned by the *Philosophical Library Company* of Salem. This library had been that of the Irish scholar Richard Kirwan, eventually finding its way to Salem—after being captured from a British ship by an American privateer—thanks to the efforts of John Prince, Joseph Willard, and other philosophically minded men. Bowditch took full advantage of the bounty, spending endless hours studying and making his own hand-copies of numerous scientific books and of the mathematical papers from the *Philosophical Transactions of the Royal Society*.[11] As a boy, Bowditch exhibited

---

[9] Biographical information may be found in Robert Elton Berry, *Yankee Stargazer: The Life of Nathaniel Bowditch* (New York: McGraw Hill, 1941). Extensive discussions of Bowditch's contributions to American science are also found in John C. Greene, *American Science in the Age of Jefferson* (Ames: The Iowa State University Press, 1984) and Dirk Struik, *Yankee Science in the Making: Science and Engineering in New England from Colonial Times to the Civil War* (New York: Dover Publications, 1991). Biographical information from contemporary sources may be found in the aforementioned Pickering, "Eulogy"; Alexander Young, *A Discourse on the Life and Character of the Hon. Nathaniel Bowditch, LL.D, F.R.S.* (Boston: Charles Little and James Brown, 1838); and N.I. Bowditch, "Memoir of Nathaniel Bowditch," in volume I of Bowditch's translation of Pierre Simon LaPlace, *Mécanique Céleste*. This memoir was originally in Volume IV of Bowditch's translation, but was moved to Volume I for the 1966 edition cited above. Most of the personal papers of Nathaniel Bowditch, including his journals and personal letters, may be found in the *Bowditch Collection* at the Boston Public Library. The Phillips Library at the Peabody Essex Museum of Salem also has a small collection of Bowditch's personal correspondence.

[10] While at sea a few years later, Bowditch also became proficient in several other languages, including Spanish, Portuguese, and Italian. Later in his life, after recognizing the increased importance of German scientific scholarship, Bowditch taught himself the German language.

[11] Works found in this library and studied by Bowditch included *Chambers' Cyclopaedia*, *Emerson's Mechanics*, Hamilton's treatise on conic sections, as well as works by

his newfound education by composing an almanac and assisting in a survey of Salem.[12] It has been argued that the scientific education Bowditch received from these books was vastly superior to one received by his contemporaries at Harvard.[13]

At the age of twenty-two, Bowditch did what many boys and young men from Salem eventually did: he put out to sea in a merchant ship. While taking part in various capacities in five lengthy sea voyages, Bowditch studied from books brought along on the trip; took meticulous notes concerning the voyage; and became an expert, in fact one may say legendary, navigator.[14] Bowditch also took the opportunity on these long voyages to instruct his fellow shipmates on the art and science of navigation, until he became famous for his crews' ability to navigate a ship, from the captain all the way down to the ship's cook. In spite of the fact that Bowditch never held a teaching position, his successful efforts in teaching navigation to his mates set the stage for a long career committed to disseminating scientific and mathematical knowledge throughout the United States and the rest of the English-speaking world.

4.2 *The New American Practical Navigator*

One of Bowditch's first discoveries, and the first on which he published a paper in the *Memoirs of the American Academy of Arts*

---

s'Gravesande, Benjamin Martin, Daniel Bernoulli and many others. Greene, *American Science*, p. 146.

[12] *Ibid.*

[13] Berry, *Yankee Stargazer*, p. 24. See also my discussion of the state of American mathematical education in Chapter 2.

[14] Tales abound concerning Bowditch's navigational prowess. In one such tale, the citizens of Salem were stunned one Christmas day to find Bowditch walking down a misty Salem street, having returned from a long voyage. Bowditch had navigated his way safely into the harbor in a dense fog that no other captain would have dared to challenge. Susan W. Bowditch, *Nathaniel Bowditch, 1773–1838* (unpublished manuscript), pp. 20–21.

and Sciences, was an improved method for using lunar observations to find longitude.[15] Because chronometers were still undependable and expensive, finding a ship's location by this method was an important tool in navigation. In 1799, Bowditch used his navigation and mathematical talents to edit and correct John Hamilton Moore's *Practical Navigator*. Moore, an Englishman, had composed a guide to navigation that was so full of errors[16] that by the time Bowditch had published a third edition in 1802 it was essentially a new work and was appropriately credited to Nathaniel Bowditch as the author.

The New American Practical Navigator[17] almost instantly became an indispensable guide to navigation on the high seas. Known to seamen the world over as the "Bowditch," *The New American Practical Navigator* also contained most of the mathematics and astronomy needed by ordinary seamen to find their way by celestial navigation rather than by dead reckoning. It soon became a book that "every British seaman had to read if he hoped to get ahead

---

[15] Nathaniel Bowditch, "New Method of Working Lunar Observations," *Memoirs of the American Academy of Arts and Sciences*, 1804, 2:1-11.

[16] By one count, the number of errors in Moore's work numbered over eight thousand. Bowditch also added new information and reorganized large parts of the book. Raymond Clare Archibald, "The Scientific Achievements of Nathaniel Bowditch," in *A Catalogue of a Special Exhibition of Manuscripts, Books, Portraits, and Personal Relics of Nathaniel Bowditch (1773-1838)* (Salem, MA: Peabody Museum, 1937), p. 7.

[17] The full title was *The New American Practical Navigator: being an Epitome of Navigation; containing all the Tables necessary to be used with the Nautical Almanac, in determining the Latitude; and the Longitude by lunar observations; and keeping a complete reckoning at sea: illustrated by proper rules and examples: the whole exemplified in a Journal kept from Boston to Medeira, in which all the rules of navigation are introduced: also the demonstration of the most useful Rules of Trigonometry: with many useful Problems in Mensuration, Surveying, and Gauging, and a Dictionary of Sea-Terms; with the Manner of performing the most common Evolutions at sea. To which are added, some general instructions and information to Merchants, Masters of Vessels, and others concerned in Navigation, relative to Maritime Laws and Mercantile customs. From the best authorities: Enriched with a number of New Tables, with original improvements and additions, and a large variety of new and important matter: also, many thousand errors are corrected, which appeared in the best systems of navigation yet published.*

of the Yankee skippers."[18] In this monumental work, Bowditch added introductions to basic mathematics like arithmetic, geometry, trigonometry, and the use of logarithms. He also supplied instruction in astronomy, geography, and the basics of navigation. In a foreshadowing of his future career as an actuary, Bowditch included excerpts on marine insurance, adapted to American laws, from a London publication called the *Ship-Masters Assistant*. This discussion on insurance was primarily a legal explanation of insurance policies, not a mathematical evaluation of actuarial questions.

The publication of *The New American Practical Navigator* brought to Bowditch an international reputation that few American authors had achieved. An admiring member of the Royal Society called it "the best book on that subject which has ever fallen into my hands."[19] After going through numerous editions, *The New American Practical Navigator* continues to be used today. In fact, the 2002 edition may be purchased for around $125. Reviews of Bowditch's classic work on Amazon.com include comments like "as a first year student in a marine program in Canada I don't know where I would be with out 'Bowditch'", "One will find this book in every chart room on every U.S. Capital ship, Coast Guard and Navy alike," and "This is the most comprehensive book on navigation ever written." Bowditch's *New American Practical Navigator* became a timeless classic in navigation.

Three aspects of the impact of *The New American Practical Navigator* deserve mention. First, as we would expect in early nineteenth century America, this contribution to science was important for its utility. No new theories or laws of nature were propounded, and no new conceptual advances were made. Bowditch created, as the title clearly states, a practical guide. Secondly, the work

---

[18] Van Wyck Brooks, *The Flowering of New England (1815–1865)* (New York: Random House, 1936), p.51.

[19] Letter from Stephen Lee, FRS, to William Vaughan (undated), in *Bowditch Collection*.

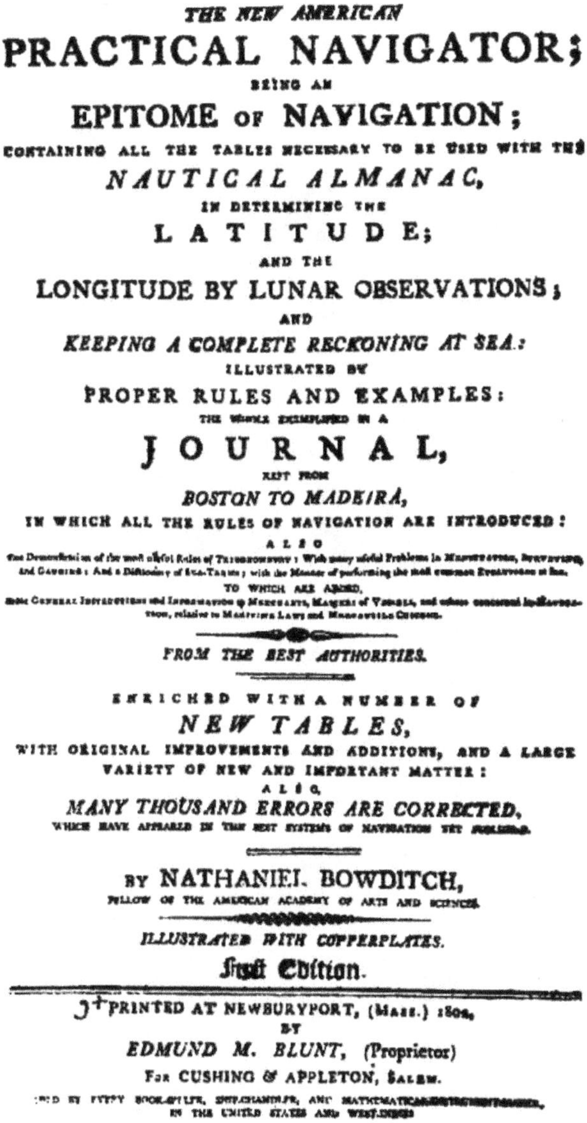

THE NEW AMERICAN
PRACTICAL NAVIGATOR;
BEING AN
EPITOME OF NAVIGATION;
CONTAINING ALL THE TABLES NECESSARY TO BE USED WITH THE
NAUTICAL ALMANAC,
IN DETERMINING THE
LATITUDE;
AND THE
LONGITUDE BY LUNAR OBSERVATIONS;
AND
KEEPING A COMPLETE RECKONING AT SEA:
ILLUSTRATED BY
PROPER RULES AND EXAMPLES:
THE WHOLE EXEMPLIFIED IN A
JOURNAL,
KEPT FROM
BOSTON TO MADEIRA,
IN WHICH ALL THE RULES OF NAVIGATION ARE INTRODUCED:
ALSO
The Demonstration of the most useful Rules of Trigonometry; With many useful Problems in Mensuration, Surveying, and Gauging: And a Dictionary of Sea-Terms; with the Manner of performing the most common Evolutions at Sea.
TO WHICH ARE ADDED,
Some General Instructions and Information to Merchants, Masters of Vessels, and others concerned in Navigation, relative to Maritime Laws and Mercantile Customs.

FROM THE BEST AUTHORITIES.

ENRICHED WITH A NUMBER OF
NEW TABLES,
WITH ORIGINAL IMPROVEMENTS AND ADDITIONS, AND A LARGE
VARIETY OF NEW AND IMPORTANT MATTER:
ALSO
MANY THOUSAND ERRORS ARE CORRECTED,
WHICH HAVE APPEARED IN THE BEST SYSTEMS OF NAVIGATION YET PUBLISHED.

BY NATHANIEL BOWDITCH,
FELLOW OF THE AMERICAN ACADEMY OF ARTS AND SCIENCES.

ILLUSTRATED WITH COPPERPLATES.

First Edition.

PRINTED AT NEWBURYPORT, (Mass.) 1802,
BY
EDMUND M. BLUNT, (Proprietor)
For CUSHING & APPLETON, Salem.
SOLD BY EVERY BOOKSELLER, SHIP-CHANDLER, AND MATHEMATICAL INSTRUMENT MAKER,
IN THE UNITED STATES AND WEST-INDIES.

Figure 4.2.1: Title page from Bowditch's *New American Practical Navigator* (1802)

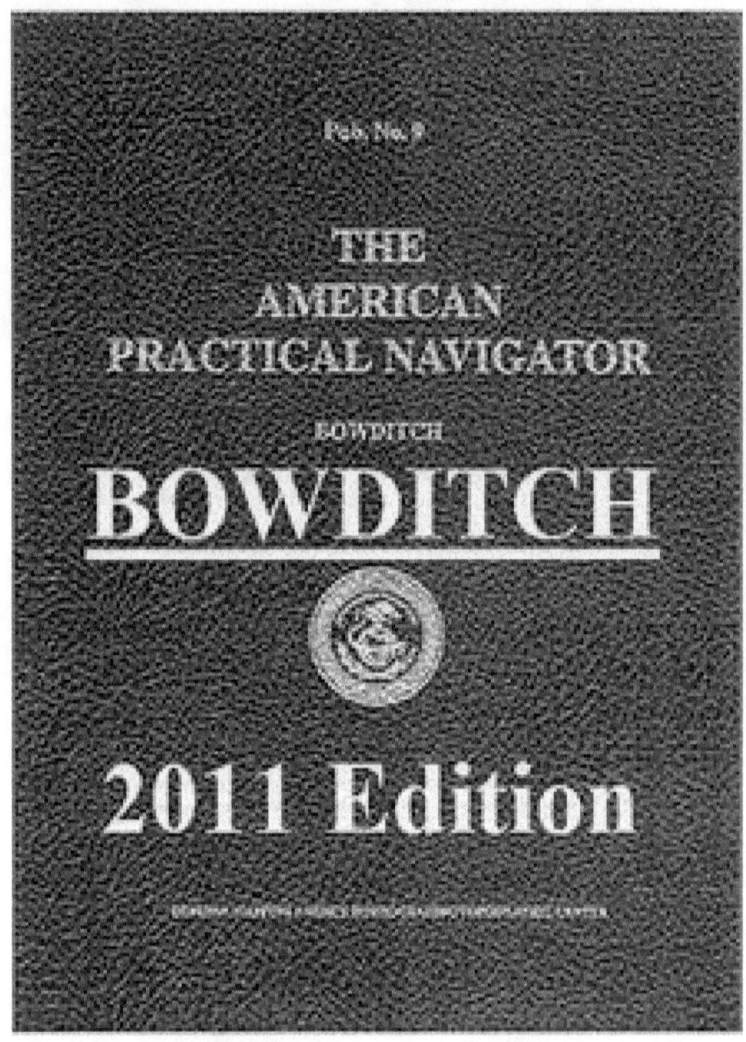

Figure 4.2.2: Cover of the Bicentennial Edition of Bowditch's *American Practical Navigator* (2002)

was initially based on a British publication. Despite the fact that voluminous changes made by Bowditch earned him the right to place his name as the author, at a basic level this was not an original work, but rather a derivative one. Finally, much of Bowditch's work on the *Practical Navigator* was done for educational purposes. Bowditch actively participated in the pursuit of educating Americans in science. This dedication to education was essential if the developing nation was to catch up with, and eventually compete with, European science.

The year 1802 was especially significant for Nathaniel Bowditch. It was in this year that the *Practical Navigator* was credited to him as author. The importance of this work was not lost on the academic community, for in 1802 Bowditch was given an honorary M.A. from Harvard. It was also in 1802 that Bowditch purchased the first volume of Pierre Simon Laplace's monumental work, *Mécanique Céleste*. Bowditch acquired this important and very difficult treatise just before he was to embark on his last voyage, this time as master of the ship. This position actually required very little of Bowditch, and thus he was able to spend long calm days at sea studying his new acquisition.

After his fifth and last sea voyage, Bowditch was offered, and accepted, the position of president of the Essex Fire and Insurance Company in Salem. Bowditch spent the rest of his life as an insurance company officer, first in Salem and later in Boston. Through it all, he continued his habit of study and research after business hours. Bowditch's reputation was such that, by 1806, he was offered the Hollis Professorship of Mathematics and Natural Philosophy at Harvard. He turned this position down,[20] as he was

---

[20]It seems that turning down positions in academia in favor of business endeavors was not uncommon. John Pickering refused the chair of Greek and Hebrew at Harvard in order to continue his busy law practice. Brooks, *Flowering*, p. 50.

later to turn down offers from the University of Virginia[21] in 1818 and West Point in 1820.

The actual reasons for Bowditch's refusals to accept academic positions are not clear. In a letter dated July 8, 1806,[22] Bowditch indicated that he believed that there were others more qualified than he for the Hollis Chair at Harvard. The tone of the letter also implied that Bowditch felt a bit intimidated at the thought of following Samuel Webber, past holder of the Hollis Chair and currently president of Harvard. He may have worried that his lack of formal education would make him an outsider at Harvard, or he may have had a fear of talking in front of groups of people. In his response to Jefferson's invitation to come to Virginia, Bowditch provided several reasons for turning down the offer, the poor health of his wife being the most telling.[23]

Interestingly, Bowditch was not the only American intellectual whom Jefferson failed to lure to Virginia. Bowditch's fellow New Englander, George Ticknor, also turned down Jefferson's offers, prompting Jefferson to begin a search for qualified faculty in Europe. Jefferson wrote to Madison concerning Francis Walker Gilmer, his agent sent to Europe to find professors for the University of Virginia:

---

[21] Bowditch received a personal invitation from Jefferson to accept the position at Virginia. In a letter (Oct. 26, 1818) that can only be labeled a "sales job," Jefferson explained the philosophy of the college, boasted of the weather in Virginia, the generous salary being offered ($2000 per year), and emphasized that the college was looking for men of the first rank. *Bowditch Collection.*

[22] *Bowditch Collection.*

[23] Evidently Jefferson tried more than once to change Bowditch's mind. In a letter from Jefferson to James Madison (July 7, 1819), his collaborator in the founding of the University of Virginia, Jefferson notes that he will again try to convince Bowditch to come to Virginia, "but with not much hope of succeeding." Quoted in James Morton Smith, ed., *The Republic of Letters: The Correspondence between Thomas Jefferson and James Madison, 1776–1826* (New York: W.W. Norton and Co., 1995), p. 1813.

> I think therefore he had better bring the best he can get. They will be preferable to secondaries of our country because the stature of these is known, whereas those he would bring would be unknown, and would be readily imagined to be of the high grade we have calculated on.[24]

Note that Jefferson implied that professors from Europe would automatically be considered of higher quality than Americans. Men such as Bowditch and Ticknor represented the exception to this rule.

Whatever his stated reasons for turning down academic positions, "American universities never strongly tempted Bowditch," according to Nathan Reingold "for they had little to offer a man of his caliber."[25] American colleges did not offer a man like Bowditch prestige, monetary rewards, time for research, or any of the other benefits of a modern university. The third criterion for the emergence of a research community, ample financial support, was not to be found at American colleges during Bowditch's lifetime. Bowditch remained in the business world for the rest of his life, content to spend his leisure time in pursuit of science.

Bowditch found continued success both in his business endeavors and his scientific "hobbies" throughout his lifetime. By 1810, he had attained a standing in the community that prompted the president of Harvard, John Kirkland, to appoint Bowditch as an overseer at Harvard in the hopes that his presence would "stimulate and gratify those of the students who may be disposed to cultivate Mathematical knowledge."[26] In 1823 he left Salem for Boston and assumed the position of actuary for the Massachusetts Hospital

---

[24] Jefferson to Madison (October 6, 1824) in *Ibid.*, p.1886.

[25] Nathan Reingold, "Nathaniel Bowditch," in Charles Coulston Gillispie, ed., *Dictionary of Scientific Biography,* vol. 2 (New York: Scribner, 1981), pp. 368–369.

[26] Letter from John Thornton Kirkland to Benjamin Pickerman, May 24, 1810, in the *Bowditch Collection.* The board of overseers included government and church officials as well as other prominent members of the community. Although the composition of the board

Life Insurance Company.[27] In 1826, Bowditch was elected to the corporation of Harvard, a group of seven men who controlled the Cambridge college. It was in this official capacity that Bowditch met a young friend of his son, and a fellow student at Harvard, Benjamin Peirce. Peirce would one day become what many recognized as the first true research mathematician produced by the United States. In the meantime, Bowditch found an important outlet for his creative activities when he began publishing in the *Memoirs of the American Academy of Arts and Sciences.*

4.3 Bowditch and the American Academy of Arts and Sciences

Much has been made, and rightly so, of Benjamin Franklin's involvement in the founding and development of the American Philosophical Society in Philadelphia. A sister society, however, founded in Boston only eleven years later, receives comparatively less attention from historians.[28] This society, the American Academy of Arts and Sciences, played an important role in the development of American science, especially astronomy, by providing a forum for dissemination of astronomical observations, calculations, and theories by American scientists.

---

varied, at the time Bowditch was appointed the board included 15 members of the clergy and 15 lay members. For data relating to the board of overseers see *Historical Register of Harvard University, 1636–1936* (Cambridge: Harvard University Press, 1937).

[27] Terence O' Donnel calls Bowditch "the first American actuary." *The History of Life Insurance in Its Formative Years* (Chicago: American Conservation Company, 1936), p. 433.

[28] See Greene, *American Science,* for a discussion of the importance of the American Academy of Arts and Sciences to American science.

Although Nathaniel Bowditch was not an original founder of the American Academy of Arts and Sciences (AAAS),[29] as Franklin had been for the American Philosophical Society, he did play an instrumental role in the AAAS for the period 1804–1838, when the society was still in its infancy.[30] In fact, Bowditch's work accounts for more than one-quarter of the total number of articles published by the AAAS in volumes III and IV, covering the period 1809–1820.[31] These works helped to establish the *Memoirs of the American Academy of Arts and Sciences* as an important repository of American science and, along with Bowditch's other works,[32] played an important role in communicating American interests in science to their European contemporaries.[33]

The American Academy of Arts and Sciences was incorporated in Boston in 1780. Its founders hoped to build a society in Boston patterned after the Royal Academy of Sciences in Paris, and rivaling the American Philosophical Society in Philadelphia, with the intent to "promote most branches of knowledge advantageous to a

---

[29] Bowditch was elected to the American Academy of Arts and Sciences in 1799, nineteen years after its founding, and before he had published anything. His election may be attributed to the fact that several members of the Academy were Salem men who had some knowledge of the level of Bowditch's self-education and the talent that he possessed.

[30] Brooks refers to Bowditch as the "second Benjamin Franklin." *Flowering*, p. 51.

[31] For further discussion of other journals important to the development of mathematics in the United States, see Chapter 5.

[32] A complete list of Bowditch's publications may be found in several places, including Archibald, "Scientific Achievements," pp. 11–16.

[33] The American Academy of Arts and Sciences, like other societies in America and Europe, elected foreign scientists to its membership. Although circulation data for the *Memoirs of the American Academy of Arts and Sciences* are not available, the list of foreign members indicates a probable widespread familiarity with the publication. These foreign members included British scientists such as Joseph Priestley, Joseph Banks, William Herschel, James Hutton, Edward Jenner and Nevil Maskelyne; Frenchmen Jean d'Alembert, Georges-Louis Buffon and Marie Condorcet; the German Johann Blumenbach; and the Swiss mathematician Leonhard Euler.

community."[34] The first volume of the *Memoirs of the American Academy of Arts and Sciences* was published in 1785. Its contents were separated into three parts. Part I was composed of astronomical and mathematical papers, part II contained physical papers, and part III was for medical papers. Even though the majority of these articles were admittedly not original research, the AAAS chose to publish papers containing "many observations not entirely new"[35] for two reasons. The first was to expose American readers to scientific knowledge not readily available due to poor communications with Europe. The second reason was the belief that repetition might make these ideas "more forcibly impressed upon the mind." Of course, a third, unstated reason should be considered. The *Memoirs of the American Academy of Arts and Sciences* did not contain significant original research because very little of such research was being conducted in the United States, especially in the physical sciences. In particular, it was admitted that:

> The astronomical and mathematical papers, in this volume, will, perhaps, be the least entertaining of any in the collection, and will have the smallest number of readers. However, they are useful in such a work. Few, if any of them, contain deep speculations and obtruse researches and calculations; but they are chiefly of a practical kind.[36]

This preface to a new journal on the scene in American science denotes two of the primary concerns held by scientists in America. One was that American science was of a practical nature, and the second was that publications should help to educate the American public in science. For whatever the reasons, stated or unstated,

---

[34] *Memoirs of the American Academy of Arts and Sciences*, 1785, *1*:iv.

[35] *Ibid.*, p. ix.

[36] *Ibid.*, p. viii.

original research was not high on the agenda of the new Academy. Yet, the founding of societies such as the American Academy of Arts and Sciences, along with the publication of their journals, represents an important first step towards the professionalization of science.

The second volume of the *Memoirs of the American Academy of Arts and Sciences* was published in 1793, eight years after the first. This lengthy period between volumes was typical. From 1785 to 1833 a total of eight tomes were published, an average of one every six years. The Academy's sister society, the American Philosophical Society, exhibited a similar timeframe for the publication of its *Transactions*. Although these periodicals mark an important beginning for American science, the scientific community was not yet able to support regularly published journals.

Nathaniel Bowditch's first paper in the *Memoirs* appeared in Volume II, Part 2 (1804). This paper, entitled "New Method of working a Lunar Observation," pertained to navigational aids using lunar observations that Bowditch had earlier published in *The New American Practical Navigator*. Although not particularly exciting as an example of original mathematics, Bowditch's first contribution was, as his subsequent contributions would also prove to be, more advanced mathematically than most of the other papers published in the *Memoirs*.

It is in Volumes III and IV that Bowditch's influence on and importance to the American Academy of Arts and Sciences began to take shape. In Volume III, Part 1 (1809), four of the thirty-one papers that appeared were due to Bowditch. Of the remaining twenty-seven articles, only three pertained to mathematics or astronomy. Although Bowditch's work on his translation of Laplace's *Mécanique Céleste* was not to begin for another half-dozen years, Bowditch noted in one of his papers that he had calculated the orbit of the comet of 1807 using a method given by Laplace in *Mécanique Céleste*.[37] Bowditch's mastery of this subject was an accomplishment few other Americans of this period could claim.

The next volume of the *Memoirs* to appear was Part 2 of Volume III in 1815. Of the twenty-three papers published in this volume, seven were from the pen of Bowditch. In his various contributions to this volume, Bowditch continued to reveal his mathematical talents. He often used the methods of Laplace, made use of calculus in the form of Newton's fluxions, and even solved systems of differential equations.

It was in this volume that one of Bowditch's most original contributions to mathematics is found. Motivated by work done by James Dean at the University of Vermont, Bowditch wrote, "On the Motion of a Pendulum Suspended From Two Points," an article in which he introduced a new class of sine curves. In this paper, Bowditch demonstrated the full range of his mathematical abilities. In his mathematical analysis of the motion of a pendulum, he solved a system of second-order differential equations using "the same method that La Place has used for a single pendulum in...his 'Mécanique Céleste'."[38] Bowditch's twenty-three page analysis of the pendulum's motion concluded with a description of

---

[37] Greene calls the appearance of this comet, along with another that appeared in the skies over the United States in 1811, Bowditch's "best opportunity to display his command of celestial mechanics." Bowditch certainly made the most of his opportunity. *American Science*, p. 151.

[38] Nathaniel Bowditch, "On the Motion of a Pendulum Suspended From Two Points," *Memoirs of the American Academy of Arts and* Sciences, 1815, 3:415. The system of differential equations Bowditch solves to find the curves that describe the motion of the pendulum is:

$$0 = d \cdot \frac{dx}{dt} + \frac{\lambda x dt}{\sqrt{rr-xx}}$$
$$0 = d \cdot \frac{dy}{dt} + \frac{\lambda y dt}{\sqrt{zz+yy}}$$
$$0 = d \cdot \frac{dz}{dt} + \frac{\lambda z dt}{\sqrt{zz+yy}} - gdt$$

He makes use of a special substitution from Laplace to find the solution. Interestingly, although Bowditch uses the language of fluxions to describe his work, he employs differential notation in the actual mathematics of his paper. He even intermingles the two competing versions of calculus in the same sentence when he writes of "forces being multiplied by the fluxion of the time dt." *Ibid.*, p. 414.

a few experiments he performed in order to check his theory against observation.

Later in the century, around 1850, the class of curves described by Bowditch in his paper was rediscovered by and named after the French physicist Jules Antoine Lissajous.[39] Lissajous applied these curves, first generated by Bowditch to study the periodic motion of a pendulum, to the science of acoustics. Once again, Bowditch's contributions to the *Memoirs* represented a mathematical sophistication not seen in the other eleven papers on mathematics and astronomy contributed by other authors in this volume.[40]

To say that seven of twenty-three, or thirty percent, of the articles in Part 2 of Volume III are credited to Bowditch would still understate his influence on this particular volume. In one of the articles written by Bowditch, the author used his own observations along with observations communicated to him by other scientists, to calculate the longitudes of over forty American cities.[41] Four separate articles, which were originally letters communicated to Bowditch, noted observations made of a solar eclipse that occurred on September 17, 1811 by four different observers from different points in the country. Bowditch used these observations, and many others,[42] along with data from older observations of transits of Venus

---

[39] See Greene, *American Science*, p. 153, Archibald, "Scientific Achievements," pp. 9, 13–14, and N.I. Bowditch, "Memoir," p. 49.

[40] See Appendix 2 for a complete list of authors contributing mathematical papers to the *Memoirs of the American Academy of Arts and Sciences* between 1785 and 1833.

[41] Improving the accuracy of the determination of latitudes and longitudes of American locations was somewhat of a passion for Bowditch. Bowditch observed an eclipse of the sun from his garden on two occasions, one in 1806 and another in 1811. He used his observations, along with those made all over the United States and relayed to him, to calculate the longitudes of many locations from which the eclipse had been observed. Greene, *American Science*, p. 147.

[42] Among the numerous observations employed by Bowditch for his calculations was one communicated to him by Thomas Jefferson from Monticello. Almost four years later, Bowditch received a letter from Jefferson thanking him for the pamphlets showing longitudes of places in the United States. In this letter, Jefferson stated, "I am happy indeed to find that

and Mercury, to perform his calculations. His desire for accuracy led him to note, "most of the preceding calculations have been made in two different ways, to verify the accuracy of the results."[43] The significance of Bowditch's calculations of the longitudes of American cities is not due to its level of sophistication. It does reflect, however, the central role that Bowditch played in American science in the first few decades of the nineteenth century. Bowditch was entrusted with the data from observations painstakingly taken from points around the country. More importantly, it was Bowditch on whom American scientists relied to make the calculations necessary to turn these observations into a scientific paper for the *Memoirs* of the Academy.

In addition to the communications regarding the solar eclipse, two other articles in Part 2 of Volume III were communicated originally to Bowditch. This means that of the twenty-three articles in the volume, thirteen directly involved Nathaniel Bowditch. This all occurred some fourteen years before he became president of the Academy; therefore, the trust given to Bowditch by his contemporaries in America was earned by reputation, not given by virtue of position.

Further analysis of Volume III, Part 2, points to the importance of the *Memoirs* of the AAAS to the small but growing group of mathematicians and astronomers in the United States. In addition to the seven papers by Bowditch, there were three other articles on astronomy from authors not associated with colleges. Also, there were a total of eight articles by three different authors who held the title of Professor of Mathematics and Natural Philosophy at their respective colleges. Four of these were by James Dean at the

---

this most sublime of all sciences is so eminently cultivated by you, and that our Rittenhouse was not the only meteor of the hemisphere in which he lived." From a letter dated May 2, 1815 in the *Bowditch Collection*.

[43] Bowditch, *Memoirs*, 1815, *3*:296.

University of Vermont, three by John Farrar at Harvard College, and one by Parker Cleaveland at Bowdoin College. Although these contributions were, for the most part, simply records of observations and mathematically insignificant, the *Memoirs* provided an outlet for those scholars wishing to pursue investigations in mathematics and astronomy. Lacking such an outlet, little motivation could exist for continued work in mathematics. The result would be to discourage young scholars from pursuing such studies. So, although the level of "research" was certainly not high by European standards, the *Memoirs* of the AAAS were crucial to the foundation being formed for an American mathematical community.

Volume IV of the *Memoirs of the American Academy of Arts and Sciences* was published in two parts, with Part 1 appearing in 1818 and Part 2 in 1820. In Part 1, seven of the sixteen total papers were by Bowditch, with another four papers from other authors on mathematics and astronomy. Part 2, published in 1820, contained a total of eleven papers, seven of which were Bowditch's. Three of the remaining contributions were of a mathematical or astronomical nature, one of which was in the form of a communication to Bowditch from A.M. Fisher, Professor of Mathematics and Natural Philosophy at Yale College.

During this period of time, 1809–1820, covering four separate publications of the *Memoirs*, slightly more than twenty five percent of the total contributions were the work of Nathaniel Bowditch. This represents far and away the most work by one author in the *Memoirs* during this period. Furthermore, the many articles that appeared in the *Memoirs* that had originated as communications to Bowditch give further evidence to the importance of Bowditch to the Academy. Interestingly, many of these communications came from Professors of Mathematics and Natural Philosophy at the leading American colleges of the time. These letters, addressed to an insurance businessman in Salem, Massachusetts, indicate the relative lack of the standing of science in academia in early nineteenth-century America. By the end of the century, there would be curious amateurs hopefully submitting their ideas and observations to distinguished professors at Harvard, Yale, and other prestigious colleges. But for a time in American sciences, the tables were turned and it was the "amateur" Bowditch who was the acknowledged leader in American science.

The types of papers submitted by Bowditch to the *Memoirs of the American Academy of Arts and Sciences* demonstrate his mastery of the different subjects in which his mathematical talents were utilized. In the four publications of the Academy from 1809 to 1820, there appeared twelve essays from Bowditch on observational or mathematical astronomy, seven on surveying or navigation, and three on mechanics or electricity. Some of these works were astronomical or meteorological observations with a minimum amount of calculations attached, but many were mathematically dense. Bowditch was a fastidious calculator, whether performing calculations based on his own observations and data or correcting mistakes he found in works of Laplace, Newton, Poisson, and others.

This fastidiousness in calculations, his precision in observations, and the patience Bowditch exhibited in the long road to publishing his translation of Laplace's *Mécanique Céleste* made the sailor-turned-scientist an oddity in nineteenth-century America. Alexis de Tocqueville observed that quick movement from "men of action" was of the utmost importance in the constantly changing social order of the American democracy. It was de Tocqueville's opinion that "the darting speed of a quick, superficial mind is at a premium [in America], while slow, deep thought is excessively undervalued."[44] Bowditch's probing mind was the antithesis of

---

[44] Alexis de Tocqueville, *Democracy in America* (New York: Harper and Row, 1988), p. 461.

de Tocqueville's analysis of Americans; yet both intellectuals and common seamen valued his work.

The next volume of the *Memoirs of the American Academy of Arts and Sciences* did not appear until 1833. By this time, Bowditch was president of the Academy and chose to call this publication Volume I, New Series, because, in Bowditch's opinion, Volume I of the old series contained inferior mathematical papers. Bowditch believed that papers in the previous publications of the Academy indicated the "low state" of American mathematics in the late eighteenth century. He did not want to send foreign members of the Academy copies of any volumes in the old series, even though all of his own contributions occurred there. It is thus rather surprising that very little in the way of mathematics or astronomy occurred in this, the first volume in the new series. There did appear, however, three papers on actuarial science contributed by Bowditch's son, J. Ingersoll Bowditch, and probably reflecting Nathaniel Bowditch's own interest in the mathematics of insurance.

Bowditch's attitude towards earlier volumes of the *Memoirs* reflects the mood of scientists and other intellectuals in the United States at the time. As was discussed in Chapter 2, the War of 1812 greatly increased the patriotic feelings of Americans, and these feelings were instrumental in the changing attitudes towards science. No longer satisfied with being the scientific doormats to Europe, American scientists, like American poets, historians, and other intellectuals, sought to break away from their colonial past and open a future in which American contributions to the arts and sciences were equal to those made by Europeans.[45]

---

[45] See, for instance, Stanley M. Guralnick, *Science and the Ante-Bellum American College* (Philadelphia: The American Philosophical Society, 1975); George H. Daniels, *American Science in the Age of Jackson* (New York: Columbia University Press, 1968); Russel Blaine Nye, *The Cultural Life of the New Nation* (New York: Harper and Row, 1960); and Donald R. Hickey, *The War of 1812* (Urbana and Chicago: University of Illinois Press, 1989).

4.4 Translation of Laplace's *Mécanique Céleste*

Whereas the *New American Practical Navigator* established Bowditch's reputation as a practical man of science, and the *Memoirs of the American Academy of Arts and Sciences* provided an outlet for his astronomical observations and calculations, it was his translation of and commentary on Laplace's *Mécanique Céleste*[46] that had the greatest impact on Bowditch's scientific reputation, both in America and in Europe.[47] This work was especially important for its effect on American science and on the European perception of American science.

Although Bowditch performed most of the translation of *Mécanique Céleste* in the years 1815–1817, the four volumes[48] did not begin to appear in print until 1829, with the fourth volume finally being published in 1839, shortly after Bowditch's death.[49] The

---

[46] *Mécanique Céleste* was one of the most important scientific works of the post-Newtonian era. It helped to bring the full power of Newton's *Principia* to the world. Although Laplace's style was difficult even for accomplished scientists to comprehend, the examples and explanations found in *Mécanique Céleste*, written in terms of the continental style of analytic mathematics instead of Newton's synthetic style, permitted a more complete appreciation of the mathematical mechanics of the *Principia*. See I. Bernard Cohen, *Franklin and Newton: An Inquiry into the Speculative Newtonian Experimental Science and Franklin's Work in Electricity as an Example Thereof* (Philadelphia: The American Philosophical Society, 1956). Bowditch called Laplace's *Mécanique Céleste* a work designed "to reduce all the known phenomena of the system of the world to the law of gravity, by strict mathematical principles; and to complete the investigations of the motions of the planets, satellites, and comets, begun by Newton in his Principia." Quoted in Greene, *American Science*, p. 155.

[47] Even before his translation of Laplace was published, Bowditch had been elected as a corresponding or honorary member of the Royal Society of Edinburgh (1818), the Royal Society of London (1818) and the Royal Irish Academy (1819). After the appearance of the translation, Bowditch was similarly honored by the Royal Astronomical Society (1832), the Accademia delle Scienze e belle Lettere di Palermo (1835), and the British Association for the Advancement of Science (1835). Archibald, "Scientific Achievements," p. 10.

[48] Laplace's *Mécanique Céleste* was a five-volume work. The fifth volume was not published until the year of Laplace's death, 1827. Although Bowditch was familiar with and even referred to the fifth volume in his commentary, he translated only the first four volumes.

[49] Much like the *New American Practical Navigator*, Bowditch's edition of *Mécanique Céleste* has found incredible longevity. Recently, in the American Mathematical Society's advertisement for its edition of the work, *Nature* is quoted: "Much more than a translation;

long delay was due to several factors. Bowditch, in his introduction to Volume I, cites his expectation that the author (Laplace) would publish a new edition with corrections and additions to the original. Although Bowditch delayed publication of his translation in deference to the original author, Laplace's corrections and additions never appeared. Later, Bowditch became reluctant to accept any financial assistance for the publication, preferring to wait until he could afford to bear the cost of printing himself.

The American Academy of Arts and Sciences made offers to help defer the cost of publication of *Mécanique Céleste*, and numerous subscription applications were made to Bowditch for a copy of the translation. John Adams was one individual who expressed interest in such a subscription, in an 1818 letter in which he called Bowditch "among the greatest Masters of the subliminal sciences which Human Understanding is capable of comprehending."[50] In the same year, Bowditch received a plea from Benjamin Vaughan[51] to reconsider his refusal to publish *Mécanique Céleste* immediately. Vaughan pointed out that Bowditch had given much of his life to the edition and others would be willing to give some of their money for its publication. He argued that immediate publication would help America's scientific reputation and provide a better opportunity for those interested in studying celestial mechanics. In a plea strangely reminiscent of Halley encouraging Newton to publish his own discoveries, Vaughan concluded his letter by saying, "You will pardon me for wishing you to reconsider an opinion, in which you so conspicuously place yourself in opposition to the public benefit."[52]

---

indeed, the extent of Bowditch's own contributions equals, or perhaps exceeds, that of the translation proper...Bowditch's commentary restores all the intermediate steps omitted by Laplace...the notes also contain full accounts of progress subsequent to the publication of the original volumes."

[50]Letter from John Adams to Nathaniel Bowditch, 1818, in the *Bowditch Collection*.

[51]*Bowditch Collection*.

[52]*Ibid.*

In spite of all of the encouragement and offers that Bowditch received concerning the publication of his translation, he eventually spent about $12,000, representing one-third of his life savings, to complete publication of his work.[53]

In a time when European (especially British) critics usually degraded American scholarship and American science, Bowditch's translation was extremely well received by his European contemporaries. Charles Babbage declared the English translation of *Mécanique Céleste* "a proud circumstance for America that she has preceded her parent country in such an undertaking."[54] The *London Quarterly Review*, after Bowditch's publication of the first volume in 1829, called the work "invaluable" to the student of celestial mechanics.[55] The same review called

> the idea of undertaking a translation of the whole *Mécanique Céleste*... one which, from what we have hitherto had reason to conceive of the popularity and diffusion of mathematical knowledge on the opposite shores of the Atlantic, we should never have expected to have found originated—or, at least, carried into execution, in that quarter.[56]

This compliment, given in a tone of surprise at the fact that an American could prepare such a work, sheds light on the lack of esteem with which American science was held by Europeans. Another European scientist, F.X. van Zach, called Bowditch "le

---

[53] Berry, *Yankee Stargazer*, p. 211.

[54] Greene, *American Science*, p. 155.

[55] *London Quarterly Review*, 1832, *48*:558

[56] *Ibid.*

premier, et jusqu'à-présent le seul géométre en Amérique."[57] Complimenting Bowditch, while subtly (and sometimes not so subtly) demeaning mathematics in America, seemed to be a pattern for European scientists.

The surprise that an American was capable of such work is reminiscent of attitudes concerning accomplished female scientists. Note the similarities between the wording of the aforementioned review of *Mécanique Céleste* and the review of Mary Somerville's *The Mechanism of the Heavens*: "This unquestionably is one of the most remarkable works that female intellect ever produced, in any age or country; and, with respect to the present day, we hazard little in saying that Mrs. Somerville is the only individual of her sex in the world who could have written it."[58] American science, in its infancy, was viewed by Europeans in the same condescending manner as were women in science. There seemed to be astonishment and delight that either an American or a woman was capable of producing important scientific contributions.

Complimentary reviews from British sources should come as no surprise; after all, the translation was in their own language and provided an obvious benefit for any English-speaking student of science. What is surprising are the comments that came from non-English sources, even the French themselves. Lacroix wrote to Bowditch that

> I perceive, that you do not confine yourself to the mere text of your author and to the elucidations which it requires; but you subjoin the parallel passages and subsequent remarks of those geometers who have treated of the

---

[57] Zach was the editor of the *Monatliche Correspondenz*, a leading German astronomical journal. He became one of Bowditch's greatest European admirers, including several examples of Bowditch's work in his journal and generally calling the attention of his readers to the American astronomer/mathematician. Greene, *American Science*, pp. 154–155.

[58] *The Edinburgh Review*, 1821. Quoted in Young, *Discourse*, p. 51.

> same subjects; so that your work will embrace the actual state of science at the time of its publication.[59]

Further evidence of the importance of Bowditch's commentary comes from another letter from Lacroix, in which he states, "I have already had occasion to recommend it [Bowditch's translation and commentary] to a young professor at Lausanne, who requested of me some explanations of the work of La Place."[60] This is truly a remarkable statement! A French mathematician was using the English translation of a classic French text in order to come to a better understanding of the original. Bowditch's *commentary*, in many respects, was of much more lasting importance than the English *translation*.[61] Its impact was felt not just in America, but also in Europe.

A tribute from the Director of the Royal Observatory at Palermo sums up the attitude of European scientists towards Bowditch and his immense work:

> Bowditch has filled up, and in a superior manner, the design of *Mécanique Céleste*, and has, moreover, corrected certain blemishes which have been noted in that work. Those comments and those notes, in my opinion, place Bowditch at the head of living mathematicians.[62]

This is a rather surprising statement when the quality of European mathematicians is considered. Although Bowditch is not remem-

---

[59] Letter from M. Lacroix, April 5, 1830. Printed in Pickering, "Eulogy," p. lxxiv.

[60] Letter from M. Lacroix, January 18, 1833. *Ibid.*

[61] Portions of *Mécanique Céleste* had been translated previously, but not in its entirety and not with the extensive commentary supplied by Bowditch. One of the earliest partial translations was made by Rev. John Toplis in England in 1814. Toplis translated only the first of sixteen books of Laplace's work. D.M. Cannell, "George Greene: An Enigmatic Mathematician," *American Mathematical Monthly*, 1999, *106*:136–151, on p. 140.

[62] Letter from Niccolò Cacciatore, May 1, 1836. In Pickering, "Eulogy," p. lxxv.

bered today in the same class as Gauss and Cauchy, his translation was obviously valued by his contemporaries as much as any other work of the time, if not for its originality, then for its clarity and utility.

As the previous tributes have shown, Bowditch's translation of *Mécanique Céleste* not only supplied the English-speaking world with Laplace's original work, but his voluminous commentary[63] made the text accessible to countless people unable or unwilling to follow Laplace's difficult mathematics. In his introduction to the translation, Bowditch stated that his primary purpose was to reduce the time and labor required to understand Laplace's work by even "persons, who have a strong and decided taste for mathematical studies." Bowditch undoubtedly succeeded in this aim. The Council of the Royal Astronomical Society called Bowditch's translation

> ...unquestionably fitted to bring the *Mécanique Céleste* within the grasp of a number of students exceeding five times, at least, that of those who could master Laplace by themselves.[64]

Even the noted British mathematician Charles Babbage commented, "I have by its [Bowditch's translation] assistance been relieved from many an hour of weary labor which I have thus been enabled to devote to my own undertakings."[65] Clearly, the commentary added by Bowditch was an important tool utilized by student and professor alike.

---

[63] Bowditch's commentary more than doubled the length of Laplace's original work.

[64] Quoted in Greene, *American Science*, pp. 155–6.

[65] Letter to Nathaniel Bowditch's sons after their father's death, August 24, 1839. In the *Bowditch Collection*.

How difficult was Laplace's original text? It had been noted in several European publications[66] that hardly twelve persons in all of Britain could read and understand *Mécanique Céleste*. Bowditch himself stated, "Whenever I meet in Laplace with the words 'Thus, it plainly appears,' I am sure that hours, perhaps days, of hard study will alone enable me to discover *how* it plainly appears." This difficulty explains, perhaps even more than the length of the work, the many years that Bowditch spent on the project.[67]

Some examples of the insertions made by Bowditch are indicative of how Bowditch was transforming Laplace's work into a textbook of celestial mechanics. For instance, in the introduction, Bowditch supplied a list of formulas, mostly involving trigonometric functions, with which Laplace had simply assumed the reader would already be familiar. These formulas were used in various places throughout the text, both by Laplace in his original work and by Bowditch in his commentary. In other places Bowditch gave a working definition of a function,[68] defined partial and exact differ-

---

[66] *Quarterly Review*, 1832, *47*:558 and *Edinburgh Review*, 1808, *11*:281.

[67] Compare Bowditch's ability to comprehend Laplace without any formal training to Thomas Carlyle, a Scottish historian and writer who held aspirations of becoming a mathematician as a student of John Leslie. Craik writes that Carlyle was unsuccessful in his struggle to understand Laplace, primarily because "the education then available in Edinburgh (or anywhere else in Britain at the time) had failed to equip Carlyle with sufficient knowledge of the differential and integral calculus, and his talent... was not great enough for him to acquire it for himself." Alex Craik, "Geometry versus Analysis in Early 19th-Century Scotland: John Leslie, William Wallace, and John Carlyle," *Historia Mathematica*, 2000, *27*:133–163, on p. 146.

[68] "A quantity z is said to be a function of another quantity x, when it depends on it in any manner." This definition was similar to the accepted definitions given by other mathematicians of Bowditch's era. For instance, according to a 1755 paper by Leonhard Euler, "If some quantities depend on others in such a way as to undergo variation when the latter are varied, then the former are called functions of the latter." Quoted in Morris Kline, *Mathematical Thought from Ancient to Modern Times* (Oxford: Oxford University Press, 1972), p. 506. Similarly, in a definition given in 1810 by Lacroix, we find the following: "Every quantity whose value depends on one or more other quantities is called a function of these latter, whether one knows or is ignorant of what operations it is necessary to use to arrive from the latter to the first." Quoted in Victor J. Katz, *A History of Mathematics* (Reading MA: Addison-Wesley, 1998), p. 724. Bowditch is obviously not trying to break

entials, and supplied many examples using the mathematical tools needed to understand Laplace's difficult derivations.[69] An example of this is found in note 14b, page 12, where Bowditch stated:

> I shall in this and the three following notes, investigate the equations of a right line, a plane, and a spherical surface, which will frequently be wanted in the course of this work.

What followed were several pages of instruction not originally in Laplace's work, but essential to an understanding of the subsequent mathematics.

Another example, indicating the level of explanation supplied by Bowditch, is found in Book II where Laplace wrote about the attraction of spheroids. For the spherical coordinates

$$r = \sqrt{x^2 + y^2 + z^2}, \quad \cos\theta = \frac{x}{\sqrt{x^2 + y^2 + z^2}}, \quad \text{and} \quad \tan\omega = \frac{x}{y},$$

Laplace gave, without explanation, the partial derivatives

$$\frac{dr}{dx} = \cos\theta, \quad \frac{d\theta}{dx} = -\frac{\sin\theta}{r}, \quad \text{and} \quad \frac{d\omega}{dx} = 0.$$

Bowditch filled in the details of the calculation with fifteen lines of explanation in which he led the reader through all the steps to find the required partial derivatives.[70]

In addition to the desire to explain more thoroughly difficult mathematical techniques, a secondary purpose of Bowditch's commentary was to give credit to Laplace's predecessors, something

---

new ground, he is simply supplying an important definition that is absent from Laplace's original work.

[69] Bowditch began many notes pertaining to difficult passages with the words "To illustrate this by a simple example, we shall suppose."

[70] This problem is found in Bowditch's *Celestial Mechanics*, pp. 283–284.

which Laplace himself was not in the habit of doing. Bowditch's eulogist cited the commonly held complaint against Laplace:

> His [Laplace's] contemporaries in France complained, that he was not willing to be just either to them, or to his predecessors, that his great fault was, his not citing the authors to whom he was indebted; and that he permitted the discourses of others to appear to the world as his own.[71]

This was certainly not an unusual complaint about Laplace. His other great work, *Théorie Analytique des Probabilités* was also highly inaccessible for similar reasons. It lacked appropriate explanatory passages and it omitted "various historical introductions."[72] Bowditch was able to remedy both shortcomings of *Mécanique Céleste*.

Finally, Bowditch had a third reason for the commentary. He was able to update much of Laplace's work with new discoveries and new mathematical techniques that had been developed since Laplace originally published *Mécanique Céleste*. In this respect, the long publishing delay was an advantage, as it gave Bowditch the opportunity to update the treatise. These updates are found throughout the four volumes. One example of Bowditch's additions to Laplace's original work is found in a 150-page appendix to Volume III in which Bowditch gave "important improvements made by Gauss, Olbers,[73] and others" in calculations concerning the orbits of the planets. This appendix also contained many pages of

---

[71] Pickering, "Eulogy," p. lvii.

[72] Ivo Schneider, "Laplace and Thereafter: The Status of Probability Calculus in the Nineteenth Century," in Lorenz Krüger, Lorraine Daston, and Michael Heidelberger, eds. *The Probabilistic Revolution* (Cambridge, MA: The MIT Press, 1987), pp. 191–214, on p. 205.

[73] Heinrich Wilhelm Matthias Olbers (1758–1840) was a German astronomer who specialized in comets and their orbits.

formulas and tables Bowditch thought necessary for a thorough understanding of the text.

Bowditch noted that he chose to present more notes on elementary principles than might have been desirous, primarily due to the lack of student preparation commonly found in American schools. He also made recommendations to young people in their first reading of the volume, thus suggesting that Bowditch had the goal of educating American students in mind as he prepared his commentary. These helpful annotations were appreciated both by accomplished scientists and by beginning students of astronomy.

As we have seen, Bowditch's translation of and commentary on *Mécanique Céleste* was very well received by his fellow mathematicians and astronomers in Europe. He received congratulatory letters from such luminaries as Herschel, Babbage, Legendre,[74] and Bessel,[75] to name a few. But to understand the reception of Bowditch and his works by his American and European contemporaries, it is important to remember the perception of American science and scientists in the early nineteenth century. America had no scientists, at least in mathematics and the physical sciences, of the first rank. Although Bowditch was the leading mathematician in America, he realized that he was not of the same caliber as the leading European mathematicians. Bowditch's son, Henry, recorded a conversation he had with his father on this subject. Nathaniel Bowditch stated:

> Archimedes was of the same order of talent as Newton, and we honor him much; and Leibniz was equal to either

---

[74]Legendre called Bowditch's work "not merely a translation with a commentary" but rather "a new edition, augmented and improved, and such a one as might have come from the hands of the author himself...if he had been solicitously studious of being clear." Greene, *American Science,* p. 155.

[75]Bessel wrote to Bowditch "Through your labors...La Place's work is brought down to our own time. You yourself enrich this science by your own additions.", *Ibid.*

of them. Euclid was a second-rate mathematician, yet
I should like to see some of his handwriting. My order
of talent is very different from that of Laplace. Laplace
originates things which it would have been impossible
for me to have originated. Laplace was of the Newton
class, and there is the same difference between Laplace
and myself as between Archimedes and Euclid.[76]

It is especially interesting that Bowditch compared himself to
Euclid, while he compared Laplace to Newton and Archimedes.
The works of Newton and Archimedes were brilliant but very difficult for most readers, even talented ones. Euclid, on the other
hand, was more of a compiler than an originator. The *Elements*
was more a textbook than original mathematics. These traits of
compilation, explanation, and education are very much the same
traits as are seen in most of Bowditch's works.

The impact of Bowditch's translation of *Mécanique Céleste* is
interesting at many levels. First, there is the fact that the foundational aspects of celestial mechanics were now available to non-French reading students of astronomy. This impact was probably
negligible in Europe where most educated scientists were familiar
with French, but was more important in America where there was
not a global knowledge of the French language.[77] The translation

---

[76] Quoted in Berry, *Yankee Stargazer*, p. 218.

[77] There is conflicting data concerning the French-reading abilities of American college students. George M. Rosenstein Jr. maintains that "Few students [in early nineteenth-century America] could read French.", "American Calculus Textbooks of the Nineteenth Century," in Peter Duren, ed., *A Century of Mathematics in America*, vol. 3, (Providence, RI: The American Mathematical Society, 1988), pp. 77–109, on p. 78. On the other hand, an English translation of Lagrange's classic *Mécanique Analytique* only recently appeared in spite of the fact that it was published a decade before Laplace's *Mécanique Céleste*. In a review of this translation, made by Boissonnade and Vagliente, Massimo Galuzzi claims a translation of *Mécanique Analytique* was not necessary because all (European?) scientists in the eighteenth and nineteenth centuries read French. *Isis*, 1998, *89*:140–141. French was taught sporadically at American colleges and universities. Although French was introduced as early as 1779 in the curriculum at William and Mary [Lawrence S. Kaplan, *Jefferson and France: An Essay on Politics and Political Ideas* (New Haven: Yale University Press,

of *Mécanique Céleste* was just one of many translations of French mathematical and scientific works that would greatly influence science in America.[78]

Secondly, the extensive commentary that Bowditch supplied provided basic instruction in mathematical techniques required to understand Laplace's difficult text. This had an impact on Europeans as well as Americans, as many students were either incapable of understanding the principles without help or simply too intimidated to attempt such a reading.

Finally, Bowditch's translation had a symbolic impact. America, usually thought of as a backwater of scientific thought, had now produced a mathematical astronomer who had gained the respect of the European scientific community. Against such a backdrop, American students and scholars acquired the impetus needed to begin a climb into mathematical prominence. Bowditch's publication of his edition of *Mécanique Céleste* "opened [a door] through which American astronomers might enter into full participation in the science of celestial mechanics."[79] In addition to Bowditch, other mathematicians of his generation would do their part to prepare the way for an American mathematics community.

Nathaniel Bowditch's work is also symbolic of the changing face of mathematics in the United States. Bowditch first gained fame with *The New American Practical Navigator* and then cemented his reputation with his translation of *Mécanique Céleste*. During Bowditch's lifetime, American mathematicians gradually changed their focus from the British to the French. *The New American Practical Navigator* is symbolic of the British-dominated mathematics and

---

1967), p. 15], it was probably not until the second decade of the nineteenth century that an institution of higher learning, the Military Academy at West Point, made a concerted effort to insure that all of its students studied French.

[78] See Chapter 6 for more on the French influence on American mathematics.

[79] Greene, *American Science*, p. 157.

science that concerned itself primarily with application and utility that existed in the United States at the turn of the century. On the other hand, *Mécanique Céleste* symbolizes the evolution of mathematics in the United States to a French-inspired concern for theory, ultimately resulting in the United States' entry into the modern mathematical community.[80]

The influence of Bowditch's translation of *Mécanique Céleste* has been discussed throughout this work. However, Bowditch's work provided a stimulus to American mathematics even before it was published. Benjamin Peirce, then a student at Harvard, assisted Bowditch in editing a large portion of the four volumes before publication. According to one of Bowditch's biographers:

> Whenever one hundred and twenty pages were printed, Dr. Bowditch had them bound in a pamphlet form, and sent them to Professor Peirce, who, in this manner, read the work for the first time. He returned the pages with the list of errata, which were then corrected with a pen or otherwise in every copy of the whole edition.[81]

It might be said that the young Peirce cut his mathematical teeth on Laplace's difficult work.

Thus far, Bowditch's translation has been treated as "the greatest venture in scientific publication in America to that date" and "the most monumental American mathematical publication to

---

[80] Interestingly, Bowditch made a move in a direction that the American mathematical and scientific community would not go until much later in the century. After studying the works of German mathematical astronomers such as Gauss and Wilhelm Olbers, Bowditch became convinced that America should emulate the German system in science and scholarship. He wrote essays in the *North American Review* critical of American ignorance of the German advances in scientific research and decrying the lack of support offered science by the United States government. *Ibid.*, p. 154.

[81] N.I. Bowditch, "Memoir," p. 140.

1850."[82] But what does such adulation of a translation say about the position of American mathematics? Admitting the educational and nationalistic importance of Bowditch's work, it must also be emphasized that it was certainly not an original mathematical treatise. Its pedagogical importance notwithstanding, Bowditch's *Celestial Mechanics* added no new knowledge to the world of mathematics or astronomy. This speaks volumes to the fact that America in the first half of the nineteenth century was still not in any position to challenge European dominance in the physical sciences. As Bowditch himself admitted, he was not of the same creative talent as Laplace, and there were certainly no other mathematicians of greater talent than Bowditch in America.

Of course, intellectual pursuits of all kinds found themselves in the same predicament as science. In literature and the arts, America was searching for a national identity and for intellectual leaders to emerge. Brooks calls Bowditch's *Practical Navigator* and Webster's *American Dictionary* "monument[s] of New England learning" which "rose like a pair of imposing gateposts at the opening of an epoch."[83] Neither work may be considered to be from the mind of a creative genius; rather Bowditch, like Webster, was an organizer, a compiler, and an educator of the American public. The emergence of important original American literature, like American mathematics, would have to wait for future generations. At this point in the country's history, the emerging nation was satisfied, even proud, to claim members of the scientific community based on their abilities to simply read, understand, and communicate with the outstanding scientists of Europe. It would remain the challenge of future generations of American scientists to establish

---

[82]Louis Karpinski, *Bibliography of Mathematical Works Printed in America Through 1850* (New York: Arno Press, 1980), p. 305.

[83]Brooks, *Flowering*, p. 69.

a truly American research community in mathematics, astronomy and the other physical sciences.[84]

Nathaniel Bowditch left a lasting legacy on American mathematics, astronomy, and science in general. Bowditch, through his publications and correspondence, had a direct influence on the leading American scientists of the day. He encouraged and guided Benjamin Peirce, who would become the leading American mathematician of the next generation. In addition, Bowditch's works continued to influence students of mathematics, astronomy, and navigation for many generations.[85] His *New American Practical Navigator* continues to play a large role in the education and training of mariners all over the world, and his translation of *Mécanique Céleste* also retains an important place in mathematical astronomy. The scientific community in America, in its embryonic stage in the early nineteenth century, owed a significant part of its future development to Nathaniel Bowditch.

---

[84] For the development of American research communities in various disciplines, see Karen Hunger Parshall and David E. Rowe, *The Emergence of the American Mathematical Research Community, 1876–1900: J.J. Sylvester, Felix Klein, and E.H. Moore* (Providence, RI: The American Mathematical Society, 1994); Daniel Kevles, *The Physicists: a history of a scientific community in modern America* (Cambridge, MA: Harvard University Press, 1971); John Lankford, *American Astronomy: community, careers, and power, 1859–1940* (Chicago: University of Chicago Press, 1997); and John Servos, *Physical Chemistry from Ostwald to Pauling: the making of a science in America* (Princeton: Princeton University Press, 1990).

[85] Bowditch's translation of *Mécanique Céleste* was studied by American scientists such as William Ferrel, who went on to do important work in geophysical fluid dynamics, and by Daniel Kirkland, who became an important planetary astronomer. Greene, *American Science*, p. 156.

Chapter 5

Making Connections: Mathematics in American Scientific Journals Before 1840

The importance of publication opportunities to the development of a research community has been a central theme of the present study. Chapter 3 argued that a publication community began to emerge in American mathematics during the period 1800–1838. This chapter continues the argument begun in Chapter 3 by looking more closely at the journals themselves and what their existence implies about the progress of mathematics in America.

Thomas Kuhn counts the journals one reads in common with others as one of the requirements for membership in a professional community.[1] Without journals, textbooks, and other monographs, the scientist, the poet, the musician, and the historian have few options for communicating their work to the public, or, more importantly, to their colleagues.

---
[1] Thomas S. Kuhn, *The Structure of Scientific Revolutions*, 3rd edition (Chicago: University of Chicago Press, 1996), p. 177.

In order for the idea of "journals read" to form one of the foundations for the development of a professional community, there must be initial attempts to establish such journals. Successful journals will often be built on the ruins of failed enterprises. It is for this reason that the advent of periodicals, some dedicated only to mathematics—but most open to sciences, arts and literature in addition to mathematics—played a critical role in the formation of an American mathematical community. Although a true research community would not emerge in the United States until late in the century, these journals helped to lay the foundation upon which the community would build.

As was the case for most intellectual pursuits in the new nation, the appearance of scientific journals on the American scene lagged behind Europe. Such periodicals as the *Journal des Sçavans* (1665), *The Philosophical Transactions of the Royal Society* (1665) and *Acta eruditorum* (1682) carried new scientific discoveries to their European readership long before any such periodicals appeared in America. The first such American journal was the *Transactions of the American Philosophical Society*, which first appeared in 1771. All these journals, European and American, were non-specialized scientific journals. In America, as in Europe, it was left to later publishers to establish journals whose content was directed to specialized fields of scientific inquiry.

The publication of specialized journals for science and mathematics was a nineteenth-century phenomenon. As science became more specialized during the course of the nineteenth century, the new professional scientists demanded more specialized journals. The emergence of journals that targeted a particular audience such as mathematicians (or botanists, or mineralogists, etc.) was a requirement for the development of a community of specialists in mathematics (or botany, or mineralogy, etc.). George Daniels maintains, "the removal of a body of knowledge from the public domain is a necessary first step in creating a place for a society of experts."[2] As long as science was primarily published in periodicals aimed at a general—albeit educated—public, science could be pursued by the gentleman amateur.

The creation of the first general science periodicals was a first step in the process of the professionalization of science. The later appearance of journals whose contributors and readers were specialists in a scientific field was a bigger step in completing the process of forming a professional community. These processes were in their infant stage in early nineteenth century America.

No American periodicals dedicated strictly to mathematics existed until the nineteenth century. Those mathematics journals that were published in the first half of the century were without exception short-lived.[3] With the absence of a sustained avenue for specialized mathematical publications, American mathematicians were forced to turn to more general periodicals for publication.

Although some mathematical papers appeared in American periodicals that published literature, poetry, political exposition and science, three journals dedicated primarily to science reserved a special importance in early American mathematics.[4] These three publications, the *Transactions of the American Philosophical Society*, the *Memoirs of the American Academy of Arts and Sciences*, and *The American Journal of Science and Arts*, wielded extraordinary influence on the development of science in America. As few avenues for publishing scientific papers in the country existed,

---

[2] George H. Daniels, *American Science in the Age of Jackson* (New York: Columbia University Press, 1968), p.41.

[3] These journals, whose existence implies there were mathematically inclined men in America interested in building a professional community, will be discussed later in this chapter.

[4] These are also the same three journals analyzed in Chapter 3 in terms of the contributing authors. This chapter engages in a different sort of analysis based upon the journals themselves.

these three periodicals dominated the scientific scene in America well into the nineteenth century.

5.1 The *Transactions of the American Philosophical Society*

The incorporation of the American Philosophical Society represents America's first sustained attempt to emulate the learned societies of Europe, even before achieving independence as a nation. The society, patterned after the Royal Society of London,[5] published the first volume of its *Transactions* in 1771.[6] Like the *Transactions of the Royal Society*, the *Transactions of the American Philosophical Society* made no pretense of creating a "refereed journal" in the modern sense of the term. The Society did not judge the credibility of the papers submitted for publication, leaving that job to the author and the reader.

The stated purpose of the American Philosophical Society was to extend man's knowledge of natural philosophy. The knowledge that the Society claimed to target, at least at its inception, was not the esoteric and theoretical knowledge of "pure" science, but rather useful knowledge that might be utilized for the good of the

---

[5]Interestingly, although originally aligning itself with the Royal Society, by 1830 the preface of Volume III (new series) states, "The contents of this volume partly belong to the physical and partly to the moral sciences. In this the society has followed the example of several learned societies in Europe, and particularly of the Royal Academy of Berlin." The influence of the new German model of education and research was being felt throughout America, even in its oldest learned society.

[6]Several authors have addressed the influence of the American Philosophical Society and its *Transactions*, as well as the other major learned society of the period, the American Academy of Arts and Sciences. See especially John Greene, *American Science in the Age of Jefferson* (Ames: The Iowa State University Press, 1984); Ralph S. Bates, *Scientific Societies in the United States* (Cambridge, MA: The MIT Press, 1965); and Alexandra Oleson and Sanborn C. Brown, eds. *The Pursuit of Knowledge in the Early American Republic: American Scientific and Learned Societies from Colonial Times to the Civil War* (Baltimore: Johns Hopkins University Press, 1976).

American people. The American Philosophical Society maintained, "knowledge is of little use, when confined to mere speculation."[7] Furthermore, when "speculative truths are reduced to practice" in agriculture, trade and other areas that benefit from science, knowledge becomes useful. Therefore,

> the members [of the Society] propose to confine their disquisitions, principally, to such subjects as tend to the improvement of their country, and advancement of its interest and prosperity.[8] This emphasis on application in the charter of the Society extended to all sciences, including mathematics.

Although the founders of the American Philosophical Society initially claimed an almost exclusive interest in practical science, in reality pure science played a small part in the pages of the *Transactions* of the society. For instance, although the subjects of applied mathematics such as surveying, navigation, mechanics, and astronomy dominated the early volumes, a few contributors did write papers on pure (not necessarily advanced, however) mathematics. Some of these contributions are discussed below.

The American Philosophical Society initially formed six committees with responsibilities in the following areas:

1. Geography, Mathematics, Natural Philosophy, and Astronomy.

2. Medicine and Anatomy.

3. Natural History and Chymistry.

---

[7] *Transactions of the American Philosophical Society*, 1789, *1*:xvii. This is the corrected version of the first volume, originally published in 1771. The Society did not publish another volume until 1786, and then republished the first volume, with corrections, in 1789.

[8] *Ibid.*

4. Trade and Commerce

5. Mechanics and Architecture.

6. Husbandry and American Improvements.

Combined with the Society's stated purpose of pursuing useful knowledge, the classification of mathematics with geography, natural philosophy and astronomy guaranteed that pure mathematics would not play an important role in the Society's *Transactions*. Mathematics, for the most part, was looked upon as a handmaid to science, and science, as a handmaid to practical application. For instance, in an introductory essay to Volume III (1793), Dr. Nicholas Collin called for more research to benefit America. Among other subjects, Collin asked for "Physico-Mathematical enquiries" to meet the needs for improvements in agricultural machines, ship building, navigation, architecture, and surveying.

The practical bent of the American Philosophical Society, of course, had a tremendous influence from the outset upon the types of papers submitted and published in its *Transactions*. The "mathematical" content of Volume I, for instance, was composed of 19 papers on astronomy and surveying among the 48 papers published. These mathematical papers, four of which were contributed by David Rittenhouse, contain the extent of the mathematics in the first volume.

The quantity and quality of the mathematical papers[9] in the *Transactions of the American Philosophical Society* varied from issue to issue, dependent primarily upon the authors who submitted papers for publication. David Rittenhouse[10] dominated the

---

[9] Here, as in other places in this study, I use the term mathematical in the context of the nineteenth century, when such disciplines as astronomy, surveying, mechanics and even geography were classified as "mixed" mathematics. An interest in "pure" mathematics was relatively rare.

[10] Rittenhouse's work represented the best of American mathematics around the turn of the nineteenth century. He published articles in the *Transactions* covering what we might

mathematical and astronomical sections of the early volumes, but later contributions came from Samuel Williams (Harvard), Robert Patterson (Pennsylvania), Andrew Ellicott (West Point), Robert Adrain (Columbia) and other mathematically-inclined writers.[11]

An examination of the contents of the early volumes of the *Transactions* serves to show the mathematical interests in the journal, its contributors and its readers.[12] As previously stated, Volume I had among its 48 papers 19 in a section devoted to mathematics and astronomy.[13] In all, 42% of the total pages of Volume I are devoted to mathematical subjects. This represents a high-point in the mathematical content of the *Transactions*, primarily attributable to the ten articles addressing the much-awaited transit of Venus.[14]

Volume II (1786) did not divide the papers into sections as had Volume I. However, by counting the same subjects as applied or mixed mathematics, including optics, magnetism, surveying, astronomy, and meteorology, we find 11 of the 46 papers to be mathematical in nature. The mathematical papers represent 11% of

---

call "pure" mathematics such as "A Method of Finding the Sum of the Several Powers of the Sines" and "Method of Raising the Common Logarithm of any Number Immediately." He also provided the most advanced work in "mixed" mathematics with such articles as "To Determine the True Place of a Planet in an Elliptical Orbit, Directly from the Mean Anomaly, by Converging Series."

[11] Refer to Chapter 3 for details concerning these and other authors.

[12] For a breakdown of the mathematical papers in the early volumes of the *Transactions*, see Appendix 1.

[13] Once again, categorizing papers as "mathematical" is difficult, as mathematics could include a wide variety of applications. Volume I provided its own categories whereas the volumes that followed did not. I have tried to be true to the intentions of the first volume by categorizing papers in future volumes as mathematical if they addressed similar subjects as the mathematical papers in the first volume.

[14] For a discussion of the importance of the transit of Venus to American science in general, and the American Philosophical Society in particular, see Brooke Hindle, *The Pursuit of Science in Revolutionary America, 1735–1789* (Chapel Hill: University of North Carolina Press, 1956); Greene, *American Science*; Bates, *Scientific Societies*; and Oleson and Brown, *Pursuit*.

the total pages of Volume II. David Rittenhouse continued as the dominant contributor with six articles as compared to the four from Rittenhouse's pen appearing in Volume I. At this time, Rittenhouse was quite well-known as America's foremost mathematician. Appearing in Volume II was a letter written to Rittenhouse concerning meteorological observations made by James Madison. Madison also included his speculations on the cause of the Aurora Borealis.

Future volumes of the *Transactions of the Philosophical Society* included varying quantities of mathematical papers, but each did reserve some room for either mixed mathematics or pure mathematics, or both. Volume III (1793) and Volume IV, new series (1834) represent the low points in mathematical content with 3% of the total pages of each devoted to mathematics. Volumes IV (1799), V (1802), and VI (1809), along with Volumes I (1818) and III (1830) of the new series, each devoted between 13–19% of their total pages to mathematical subjects. The only volume not conforming to this range was Volume II of the new series, which appeared in 1825 and included 42% of its total pages as mathematical works. A large collection of papers contributed by Ferdinand Hassler, connected to his work with the United States Coast Survey, accounts for almost this entire amount.

The majority of papers published in the *Transactions* addressed topics in the category of "mixed" mathematics. Some of these articles contained calculations of longitude, but most were simply observational data. The few articles that contained "pure" mathematics were simple in form and content. For example, Robert Patterson of the University of Pennsylvania contributed an article on "An improved Method of projecting and measuring plane Angles,"[15] which gave a compass and straightedge construction for angles.

---

[15] *Transactions of the American Philosophical Society*, 1809, 6:29–31.

There were, however, a few articles in the *Transactions* that demonstrated a higher level of mathematical sophistication. For instance, "Observations on the Figure of the Earth"[16] by Joseph Clay used Newtonian fluxional notation while "Investigation of a Theorem, proposed by Dr. Rittenhouse, respecting the summation of the several Powers of the Sines; with its Application to the Problem of a Pendulum vibrating in circular Arcs"[17] by Owen Nulty, Professor of Mathematics at Dickinson College, made use of integral calculus and the notation of Leibniz. Robert Adrain, in many of his contributions, also exhibited a preference for Leibnizian notation.

In addition, articles such as "Solution of a General Case of the Simple Pendulum,"[18] by Eugenius Nulty and "On the Motions of Solids on Surfaces,"[19] by Henry James Anderson made use of calculus, including the solution of differential equations–quite a step up from many of the simple geometric and algebraic problems in previous volumes. Note that although these articles are not what might be termed "useful," they certainly did not fall into the category of pure mathematics. Rather each was an example of mathematics in service to a theoretical scientific subject.

Although the authors of some articles in the *Transactions* showed a proficiency in calculus, differential equations, and other advanced mathematical techniques, it certainly cannot be claimed that any of the contributors were outstanding research scientists (with the possible exception of Adrain, who will be discussed at length) who made important contributions to the world of mathematics. Nevertheless, they are representatives of the embryonic

---

[16] *Transactions*, 1802, 5:312–319.

[17] *Transactions*, 1818, *1*(new series):395–400.

[18] *Transactions*, 1825, *2*:466–477.

[19] *Transactions*, 1830, *3*:315–382.

stage of American mathematics whose first attempts to publish mathematical research in the United States, regardless of the quality and originality of their work, laid the foundation on which future generations of American mathematicians would build.

5.2 The *Memoirs of the American Academy of Arts and Sciences*

The American Academy of Arts and Sciences, incorporated in Boston in 1780, was, much like its older sibling in Philadelphia, designed to promote *useful* knowledge: "The purpose of this institution is to promote most branches of knowledge advantageous to a community."[20] The preface to the first edition of the *Memoirs of the American Academy of Arts and Sciences* was a patriotic call to all learned gentlemen in America to make useful contributions to the new country:

> ...the citizens have great opportunities and advantages for making useful experiments and improvements, whereby the interest and happiness of the rising empire may be essentially advanced.[21]

In particular, the first volume of the *Memoirs* reflected interest in two subjects that were deemed of vital importance to the new nation: natural history and astronomy. As was discussed in Chapter 2, Americans already perceived natural history not only as a science that they were uniquely positioned to explore, but also as one that should supply knowledge useful to everyday life. For instance, in the Academy's call for participation from learned Americans, it was pointed out that the natural history of natives

---

[20] *Memoirs of the American Academy of Arts and Sciences*, 1785, *1*:iv.

[21] *Ibid.*, p. xi.

("aboriginal inhabitants") might be useful in understanding more about how to live in the climate and take advantage of the soil.

In addition to natural history, the founding members of the American Academy of Arts and Sciences perceived a need for astronomical observations. Admitting that the mathematical and astronomical papers solicited by the Academy might be of interest to only a few readers, the call was made nonetheless to all "gentlemen capable of it" to submit astronomical observations.[22] The reason for the interest in astronomy was its application to geography and navigation, two sciences important to a country with an almost boundless land to explore. In addition, astronomical observations presented Americans with the opportunity to supply important data to the scientific world by virtue of their unique geographic location.[23] Reports on the natural history of the land and its flora and fauna, along with astronomical observations made throughout the country, dominated the contributions to the *Memoirs of the American Academy of Arts and Sciences* throughout the first half-century of its existence.

The emphasis on useful science remained a trademark of the *Memoirs* (as with the *Transactions of the American Philosophical Society*) well into the nineteenth century. For the papers classified as mathematical, astronomical observations continued to form the most numerous contributions. Some of these papers contained calculations, but most were simply unanalyzed data. A typical contribution came from Parker Cleaveland, professor of mathematics and natural philosophy at Bowdoin College. Cleaveland's "Observations of the eclipse of the sun,"[24] appearing in Volume III, part

---

[22]*Ibid.*, p. viii.

[23]Recall the importance of the observations, made by Americans, of the transit of Venus in 1769.

[24]*Memoirs*, 1815, *3*:247–248.

2 (1815), contained Cleaveland's observations that were later used by Nathaniel Bowditch in his own calculations.

An analysis of the contents of the first volume of the *Memoirs* serves to reinforce the stated objectives of the founders of the Academy. Few of the articles fit the category of basic science. Mixed mathematics, in the form of astronomy and geography, appeared most frequently; natural history and medical papers formed the majority of the rest of the volume.[25] In particular, this first volume devoted 37% of its pages to mathematical articles. With one exception, each of these contributions by a wide variety of Americans[26] emphasized the utility of their science.[27]

Like the *Transactions of the American Philosophical Society*, future volumes of the *Memoirs of the American Academy of Arts and Sciences* included a significant number of articles classified as mathematics.[28] Part 1 of Volume II, published in 1793, dedicated 19% of its pages to mathematics, and Part 2 (1804) included a total of 28% devoted to mathematics. Included among these articles was the first contribution to the *Memoirs* from Nathaniel Bowditch, who would become a critical contributor and future president of the Academy.

---

[25] By my count, 16 articles covered mixed mathematics, 11 concerned natural history, and 8 dealt with medicine, out of a total of 53 articles. This count is made by casting wide the net to include such things as simple astronomical observations in the mixed mathematics category and by considering such topics as earthquakes and agriculture in natural history.

[26] Four men, all closely associated with Harvard, dominated the mathematical papers in the first few volumes of the *Memoirs of the American Academy of Sciences*: Joseph Willard, president of Harvard from 1781–1784, Samuel Williams, Hollis Professor of Mathematics and Natural Philosophy from 1780–1788, James Winthrop, son of Harvard's most famous eighteenth-century scientist (John Winthrop), and Samuel Webber, Hollis Professor of Mathematics and Natural Philosophy from 1789–1804.

[27] The one exception was a short algebraic treatise, "On the Extraction of Roots," pp. 165–172, by Benjamin West of Rhode Island.

[28] See Appendix 1 for a summary of the number of mathematical papers appearing in each volume of the *Memoirs of the American Academy of Arts and Sciences*.

Volume III continued the trend with a still larger percentage of mathematical works. We find 19% of the pages of Part 1 (1809) addressing mathematical subjects while Part 2 (1815) devoted 63% of its pages to mathematics. The primary reason for this large percentage in the 1815 volume is the growing participation of Bowditch. The Boston mathematician/astronomer contributed a total of 11 papers to the two parts of Volume III, including one of his most original works, "On the motion of a pendulum suspended from two points."[29]

Nathaniel Bowditch's dominance of the *Memoirs of the American Academy of Arts and Sciences* continued with Volume IV. Mathematics occupied 33% of the pages of Part 1 (1818), with Bowditch being responsible for 7 of the 11 mathematical articles. Part 2 (1820) devoted 18% of its pages to mathematical papers, with 4 of the 5 belonging to Bowditch. Volume I of the new series, published in 1833, marked the end of the mathematical emphasis in the *Memoirs*. With Nathaniel Bowditch no longer a contributor, the number and sophistication of mathematical papers dropped dramatically. A few astronomical observations and two actuarial papers by J. Ingersoll Bowditch, Nathaniel's son, compose the extent of mathematics in the volume. Although future volumes would contain contributions from notable mathematicians such as Benjamin Peirce and Joseph Lovering, the death of Nathaniel Bowditch marked the end of the impressive run of mathematical papers in the *Memoirs*.

Just as with the papers in the *Transactions of the American Philosophical Society*, the papers appearing in the *Memoirs of the American Academy of Arts and Sciences* were important not so much for their content as for the very fact that Americans were attempting to enter the world of professional science in which publication played an important role. In addition to those already

---

[29] This paper was discussed in Chapter 4.

named, several other leaders in the embryonic American mathematical community published work in the *Memoirs*. These include James Dean, professor of mathematics and natural philosophy at the University of Vermont, and John Farrar, Hollis professor of mathematics and natural philosophy at Harvard. In addition, two Harvard men important to the next generation of mathematics in the United States, Joseph Lovering, Hollis professor of mathematics and natural philosophy, and Benjamin Peirce, Perkins professor of Astronomy and Mathematics, began publishing in the *Memoirs* in 1846.[30] These men helped to form a mathematical tradition in the United States that had not previously existed. Avenues for publication, such as the *Memoirs of the American Academy of Arts and Sciences*, provided an important outlet for these pioneers of American science.

The *Transactions of the American Philosophical Society* and the *Memoirs of the American Academy of Arts and Sciences* were critical publications for the early American republic. Three major weaknesses of these publications, however, conspired to retard the rate at which American science could grow. First, they were general science publications. It would be much later in the nineteenth century before successful specialized journals appeared in America.[31] The second weakness of both the *Transactions* and the *Memoirs* was their irregular publication. The *Transactions* were published ten times in the sixty-three years from 1771 to 1834, while the *Memoirs* were published only eight times between 1785 and 1833. It would be difficult to establish any sort of permanent scientific institutions without a regularly appearing outlet for scientific pub-

---

[30] Appendix 2 contains a full list of the authors who contributed mathematical works to the *Memoirs* before 1833.

[31] By successful journals, I mean those that experienced a long run and firmly established a place in their respective science. The later part of this chapter addresses the first attempts at establishing mathematical journals in the United States. None of these were "successful" in the sense of longevity.

lications. Finally, each publication was of a regional, rather than national, character. Authors from Pennsylvania and New York dominated the first ten volumes of the *Transactions of the American Philosophical Society*. Pennsylvania authors represent 53% of the total number of contributors, while New York authors represent 32% of the total.[32] Similarly, the volumes of the *Memoirs of the American Academy of Arts and Sciences* during the same time period (eight volumes in all) were dominated by New England authors. Sixty-seven percent of the pages of the *Memoirs* were the product of Massachusetts writers, with most of the remaining pages being the work of other New England authors. It was the last two of these defects, the infrequency and regionalization of the publications that were addressed by *The American Journal of Science and Arts*.

5.3 *The American Journal of Science and Arts*

The third important general-science periodical to appear on the American scene was in many ways different from the *Transactions of the American Philosophical Society* and the *Memoirs of the American Academy of Arts and Sciences*. When Benjamin Silliman founded *The American Journal of Science and Arts* in 1819, he did not associate his journal with a learned society. In spite of attempts otherwise, the American Philosophical Society and the American Academy of Arts and Sciences were regional institutions. The distances involved and the state of transportation in the United States would not allow otherwise. Therefore the journals published by the

---

[32] These totals represent the authors whose geographic location could be ascertained either through the *Transactions* themselves or through other references. Particularly valuable for this identification process are Charles Coulston Gillispie, ed., *The Dictionary of Scientific Biography* (New York: Scribner, 1981); *The Dictionary of American Biography* (New York: Charles Scribner's Sons, 1996); and Clark Elliott, *Biographical Dictionary of American Science: the Seventeenth Through the Nineteenth Centuries* (New York: Greenwood Press, 1990).

respective institutions, the *Transactions* and the *Memoirs*, were in many ways regional in nature also. Silliman sought to develop a journal that was:

> Not a *local*, but a *national* undertaking, its leading object is to advance the interests of this rising empire, by exciting and concentrating original American effort, both in the sciences, and in the arts, and it may with truth be said, that no Journal was ever more fully sustained by *original* communications.[33]

Silliman hoped to form a national journal of science that would be sustained by contributions from all across the nation. He was successful in this endeavor, at least as measured against the highly localized *Transactions* and *Memoirs*. Between the years 1818 and 1834, Silliman achieved a much wider geographical distribution for his contributors than had either of the publications of the nation's two primary learned societies. Although contributions from Connecticut authors did represent the highest percentage of any state, it was only 25% of the total. Another 17% came from New York, a state that had been a major contributor to the *Transactions*, while 37% of the contributions came from the New England states that had always supported the *Memoirs*.[34]

Silliman also met with success in his intent to publish the journal on a regular basis rather than the piecemeal publications of the learned societies. Between 1818 and 1834, *The American Journal of Science and Arts* was published in twenty-six volumes, often appearing more than once a year. Compare this to the *Transactions*, which appeared only four times between 1818 and 1834, and the *Memoirs*, which appeared a total of three times in the period.

---

[33] *The American Journal of Science and Arts*, 1820, 2: Preface.

[34] See Appendix 2 for a complete list of the authors who contributed mathematics to *The American Journal of Science and Arts*.

Figure 5.3.1: Benjamin Silliman

Silliman's ideas concerning his journal had their basis in scientific patriotism. Silliman pointed out that European countries already had a tradition of such journals, whereas with the exception of a few medical journals and infrequently published society memoirs, the United States had none. Silliman went on to say that there was much work to be done in American botany, mineralogy, and geology, and that foreign naturalists who "are frequently exploring our territory...convey to Europe the fruits of their researches."[35] Silliman wanted to found a journal "designed as a deposit for *original American communications*," but in the spirit of the education of Americans, one that "will contain also occasional selections from Foreign Journals, and notices of the progress of science in other countries."[36] Silliman hoped his journal would allow American naturalists to publish their research in America and in the process assist Americans who wished to stay current with science all over the world.

Although Silliman envisioned his journal would be different from the publications of the learned societies,[37] there was one important similarity. *The American Journal of Science and Arts*, like its pre-

---

[35] *Journal*, 1819, *1*:5.

[36] *Ibid.*, p. v.

[37] One way that the journal was different from the publications of the learned societies, much to Silliman's chagrin, was in the monetary support for publication. The *Transactions of the American Philosophical Society* and the *Memoirs of the American Academy of Arts and Sciences* had the luxury of a relatively large membership to help support the cost of publications. Silliman found himself literally begging for patronage for his journal, often appealing to the patriotism of his readers: "A more extended patronage is indispensable to its [the journal's] permanent establishment, and, should it fail on this ground, who can wonder if our national character should be even more severely (perhaps even more *deservedly*) reproached than ever." Vol II, No. 1 (April, 1820). Silliman's concerns eased enough that, two years later, he was to write, "A trial of four years has decided the point, that the American Public will support this Journal. Its pecuniary patronage is now such, that although not a lucrative, it is no longer a hazardous enterprize. It is now also decided, that the intellectual resources of the country are sufficient to afford an unfailing supply of valuable original communications." *American Journal*, 1822, Vol. 5. In four short years Silliman had found the two resources that he needed in sufficient supply: money and scientific contributors.

decessors, would emphasize (at least in the beginning) the utility of science. Even its original title laid claim to the idea that science was nothing if not useful: *The American Journal of Science, more especially of Mineralogy, Geology, and the Other Branches of Natural History; Including also Agriculture and the Ornamental as Well as Useful Arts.* Silliman emphasized, in his preface to the first volume, that this journal would "embrace the circle of the Physical Sciences, with their application to the Arts, and to every useful purpose."[38] Furthermore, "While Science will be cherished *for its own sake*, and with due respect for its own *inherent* dignity; it will also be employed as the *handmaid to the Arts*."[39] Although Silliman's journal[40] ultimately contained more "pure" science than its predecessors, its bread and butter would continue to be applications of science.

Silliman's phrasing of his outline for the new journal demonstrated the contrasting feelings often displayed by American scientists of the period. On one hand, many men of science were firm believers in the pursuit of science for its own sake. Men like Silliman and Nathaniel Bowditch[41] knew that Americans must embrace pure science in order to elevate American scholarship to the same level as that of the Europeans. On the other hand, American society, and therefore its science, was deeply imbued with the

---

[38] *American Journal*, 1819, 1:v.

[39] *Ibid.*

[40] *The American Journal of Science and Arts* was often referred to in print as *Silliman's Journal* because Benjamin Silliman almost single-handedly kept the journal alive in its early years. I use the phrase Silliman's journal in a more general sense.

[41] The dichotomy between pure and applied science was especially evident in Bowditch's life. His practicality as a seaman and businessman led him to publish The *New American Practical Navigator* and to spend his life in the insurance business. On the other hand, his translation of Laplace's *Mécanique Céleste* represents one of the high points in theoretical science in early America.

practical Scottish philosophy and the Baconian view of utility.[42] Alexis de Tocqueville theorized that such a practical bent towards learning was inherent in a democratic people, where utility would always come before aesthetics and beauty.[43] It was this dichotomy, a personal taste for the pursuit of pure scholarship versus the public's demand for usefulness, which many intellectuals, both in the sciences and in other endeavors, often struggled with in nineteenth-century America.

Silliman's closing remarks to the first volume of his journal addressed three common themes in early American science: patriotism, utility, and patronage. Silliman believed:

> To concentrate American efforts in science and the arts, by furnishing a Journal to record their [American scientist's] proceedings, will, in our view, not only have a direct influence in promoting the honour and prosperity of the nation as connected with its physical interests, but will also tend in no small degree to nourish and enlarge patriotism, by winning the public mind from odious asperities of party.[44]

Silliman continued his closing remarks with a call for more articles on the practical arts and for more financial support from all Americans interested in science.

---

[42] For a discussion of the influence of the Scottish Enlightenment on American science, see Hindle, *Pursuit*. For the inherent Baconianism in American science and its evolution into a particularly American philosophy of science, see Daniels, *American Science*. For a discussion of mathematics in Scotland in the early nineteenth century, much of which parallels events in the United States of the same period, see Alex D. D. Craik, "Geometry versus Analysis in Early 19th-Century Scotland: John Leslie, William Wallace, and Thomas Carlyle," *Historia Mathematica*, 2000, *27*:133–163.

[43] Alexis de Tocqueville, *Democracy in America*, (New York: Harper and Row, 1988), p. 465. Also see Chapter 10, "Why the Americans are more Concerned with the Applications than the Theory of Science." Tocqueville's ideas about intellectual pursuits in a democracy were summarized in Chapter 2 of the present work.

[44] p. 440.

Although not a mathematician himself, Silliman was interested in promoting research in all the sciences, mathematics included. In his introductory remarks to the first volume of *The American Journal of Science and Arts*, Benjamin Silliman expressed his desire to further mathematical research in the United States. He hoped that his journal could serve the need for a repository of mathematical papers produced by American mathematicians:

> The science of mathematics, both pure and mixed, can never cease to be interesting and important to man, as long as the relations of quantity shall exist, as long as ships shall traverse the ocean, as long as man shall measure the surface or heights of the earth on which he lives, or calculate the distances and examine the relations of the planets and stars; and as long as the *iron reign of war* shall demand discharge of projectiles, or the construction of complicated defences.[45]

Note that Silliman gave brief mention to pure mathematics. Rather his call for mathematical research was a call for inquiry into the applications of mathematics: mensuration, navigation, surveying, astronomy, and ballistics. Although the number of contributions to Silliman's journal on pure mathematics exceeded those found in the publications of the American Philosophical Society and the American Academy of Arts and Sciences,[46] the emphasis of the majority of mathematical contributions, as with other scientific papers, continued to be on application.[47]

---

[45] *Ibid.*, p. 8.

[46] By my count, 21 articles appearing in Silliman's *Journal* from 1818–1830 might be classified as addressing pure mathematics. By this I mean works that seek to solve mathematical problems or provide mathematical exposition with little or no view towards their application to other sciences. Every volume except the first in 1818 and the fourth in 1821 contained at least a small amount of pure mathematics.

[47] See Appendix 1 for a breakdown of the mathematical articles appearing in *The American Journal of Science and Arts* from 1819–1830.

Volume I of *The American Journal of Science and Arts* was published in four numbers in 1819. In these four publications, only one contribution was made in the field of mathematics, that being a derivation of familiar trigonometric formulas by Theodore Strong, professor of mathematics at Hamilton College. In fact, Strong[48] made the only mathematical contributions to the first two volumes of the journal (1819–1820). Each of these contributions was of a rather simplistic nature and appeared to be the type of article a struggling new journal might accept in order to fill its pages. Not surprisingly, the majority of the papers in the journal dealt with geology and mineralogy, topics close to the heart of the journal's founder and editor.

A look at the articles in these first few volumes of *The American Journal of Science and Arts* serves to reinforce the conclusion that American mathematics and physical science operated at a distinct disadvantage to natural history. For example, Volume III (1821) contained numerous papers with a distinctive American flair. Articles on topics such as "the Mineralogy and Geology of parts of South and North Carolina," "on the amber of Maryland," and "on Geology, &c. of the N.W. Portion of Lake Huron," could only have been produced in America, if not only by Americans. Many other articles on such diverse subjects as botany, entomology and fossil zoology concerned subjects that, by their very nature, were endemic to America. America's unique flora and fauna gave American naturalists an advantage in the study of their science. With the possible exception of the benefit of place in astronomical observations, no such advantage existed for Americans studying math-

---

[48] Strong's importance to *The Journal* was discussed in some detail in Chapter 3. Hogan believes that a series of articles contributed by Strong that employed Leibnizian notation was more influential than Nathaniel Bowditch's translation of *Mécanique Céleste* because of their wider circulation and less formidable structure. He does not, however, supply evidence for this claim. Edward R. Hogan, "Theodore Strong and Ante-Bellum American Mathematics," *Historia Mathematica*, 1981, 8:439–455, on p.447.

ematics or the physical sciences. Although there might be said to exist an "American" natural history, there was not, and really could not be, an "American" mathematics.

Beginning in Volume V (1822) and continuing through Volume XXV (1834), *The American Journal of Science and Arts* published mathematical articles in greater quantity and of much greater quality than any other periodicals of the time.[49] It was during this time that Silliman's journal became an important part of American science, and, in mathematics at least, there were no serious challengers to its primacy.[50] Although none of the articles could be said to have any permanent importance to mathematics, they did represent a level of mathematical sophistication not seen anywhere else in America up to that time. In addition, several issues in American mathematics were addressed in these articles. One of these issues was the question of the proper use and form of calculus.

American mathematicians were, much like their contemporaries in Britain, beginning to question their long-held affinity for Newtonian-style calculus, or the method of fluxions. The method of fluxions[51] essentially achieved the same results as the method of differential calculus developed by Leibniz and adopted by most

---

[49]During this period only one of the several short-lived mathematical journals was in existence. This journal, *The Mathematical Diary*, will be discussed later in this chapter. For the most complete and reliable reference work related to mathematical publications in the United States, see Louis Karpinski, *Bibliography of Mathematical Works Printed in America Through 1850* (New York: Arno Press, 1980).

[50]Later in this chapter I discuss periodicals arising in America dedicated to mathematics only. None of these periodicals reached the level of importance of *The American Journal of Science and Arts* for mathematical content in spite of their concentration on the subject of mathematics.

[51]According to Newton's definition, a fluxion is the component velocity of a continuously moving point. If x is the horizontal component of the direction of the moving point and y is the vertical component of the direction, then Newton defined $\dot{x}$ and $\dot{y}$ as the horizontal and vertical components of the motion of the point. He called these motions fluxions. In essence, a fluxion is equivalent to Leibniz's differential, and the computational (if not the mathematically rigorous) equivalent to a derivative.

European mathematicians. However, Newton's awkward notation system, along with the freedom with which continental mathematicians applied differential calculus with little regard for metaphysical foundations, resulted in the calculus of Leibniz becoming much more useful and advancing at a faster rate than that adopted by the British (and American) mathematicians who followed their master, Newton.[52]

Sometime after the turn of the nineteenth century, some British mathematicians finally began to acknowledge the superiority of the analytical methods developed in the previous century by continental mathematicians. This realization resulted in the formation of the Analytical Society in 1813 by a threesome of Cambridge undergraduates, Charles Babbage, John Herschel, and George Peacock. The formation of the Analytical Society is often seen as the beginning of the end of fluxional methods in England.[53]

It was against this background that Americans interested in mathematics tackled some of the questions surrounding the usage of calculus and its logical foundations. The pages of *The American Journal of Science and Arts* are an ideal place to trace some of the tensions arising as American mathematicians competed for authority within their small group. The mathematical papers that have

---

[52] There are numerous books covering the discovery of calculus by Newton and Leibniz, the development of calculus after its discovery, and the controversy over the rightful first discoverer. See for instance Carl Boyer, *The History of Calculus and its Conceptual Development* (New York: Dover, 1959); C.H. Edwards, Jr., *The Historical Development of the Calculus* (New York: Springer-Verlag, 1979); A. Rupert Hall, *Philosophers at War: The Quarrel Between Newton and Leibniz* (Cambridge: Cambridge University Press, 1980); and Niccoló Guicciardini, *The Development of Newtonian Calculus in Britain 1700–1800* (Cambridge: Cambridge University Press, 1989).

[53] For discussions on the development and influence of the Analytical Society, see P. J. Enros, "The Analytical Society (1812–1813): Precursor of the renewal of Cambridge mathematics," *Historia Mathematica*, 1983, *10*:24–47; M. V. Wilkes, "Herschel, Peacock, Babbage and the Development of the Cambridge Curriculum," *Notes Rec. R. Soc. Lond*, 1990, *44*:205–219; and Susan Faye Cannon, *Science in Culture: the early Victorian period* (New York: Science History Publications, 1978). For the Scottish influence on the adoption of analysis in Britain, see Craik, "Geometry versus Analysis."

something to say about calculus, either its use or its logical foundations, may be divided into two categories: those that make use of either fluxions or differential calculus without reference to why the particular method is chosen, and those papers that do make an argument for the superiority of their chosen method of doing calculus.

5.4   The Calculus Question: Differentials or Fluxions, Synthesis or Analysis?

In the first category are papers such as "On Maxima and Minima of Functions of two variable quantities"[54] by A.M. Fisher, professor of mathematics and natural philosophy at Yale. This paper is an early example of the use of differential notation at a time when American authors, like their British counterparts, continued to struggle with the debate between Newton's fluxions and Leibniz's differential calculus. Admitting that his paper did not add a new "instrument to the adept in Analysis," Fisher nonetheless hoped:

> It may still perhaps be regarded with some interest by those who are desirous of giving the greatest possible extent to the ordinary method of obtaining maxima and minima, in consequence of not enjoying the opportunity of becoming familiar with all the refinements of the modern calculus.[55]

In this passage, Fisher acknowledged there were mathematicians in America "adept in [modern] Analysis," but evidently the majority

---

[54] 1822, 5:82–93. Fisher's article was written in 1818 and originally appeared in the *Memoirs of the Connecticut Academy of Arts and Sciences*.

[55] *Ibid.*, p. 93.

of the readers of Silliman's journal had not "enjoyed the opportunity" of becoming familiar with these techniques. In this respect, Fisher's contribution may be viewed as a subtle attempt at proselytizing American scientists in the ways of modern analysis.

Fisher's paper was unusual and, it seems, a little ahead of its time. The pages of *The American Journal of Science and Arts* contained many contributions addressing the fluxion/differentials debate and the logical foundations of calculus.[56] Fisher's work, on the other hand, embraced the continental style of differential calculus without mention of the methods and notation of Newton.

Other calculus papers in Silliman's journal fall into our first category, but some of these solved problems using fluxions without much mention of its relation to differential calculus. One example of this type of paper is "Problem to determine the position of the Crank when the tendency of the power to produce the rotation is maximum" by A.B. Quinby.[57] This problem is a maximization problem using fluxions without comment on the relative merits of the method of fluxions versus differential calculus.

Numerous contributions to Silliman's journal fall into the second category, expositions on the relative merits of fluxions and differential calculus. In the years 1824 to 1830, *The American Journal of Science and Arts* published many articles attacking the foundations of analysis,[58] and others defending the fluxional methods of Isaac Newton against Leibniz's differential calculus and the newer

---

[56] A discussion of some of these papers appears below.

[57] 1826, *11*:338–339.

[58] The term analysis was used in two different ways. Sometimes it referred to calculus in particular, whether fluxions or differential calculus, and other times it was used in a more general way to mean any type of symbolic (algebraic) technique. In all cases analysis was contrasted to the synthetic (geometric) method of the ancient Greeks, a method that was the only acceptable avenue for mathematical demonstration in the eyes of many mathematicians. The difference between analysis and synthesis was particularly important to British and American mathematicians, as their acknowledged intellectual guide, Isaac Newton, composed his historic *Principia* in the synthetic style. Unless otherwise noted, I use the term analysis as synonymous with calculus. See Helena Pycior, "British Synthetic vs. French Analytical

analytical methods of Lagrange and other European mathematicians. One may find at least a half dozen articles whose primary topic was the question of fluxions versus differentials or the relative merits of the synthetic method versus the analytical method. In addition, several articles whose subjects varied over the spectrum of mathematics addressed these questions in passing. The subject of the proper arena for calculus usage was a dominant mathematical theme in *The American Journal of Science and Arts* for the better part of a decade.

An example of one who bitterly resented the use of the new methods of analysis was an anonymous reviewer of "A Treatise of Mechanics" by Olinthus Gregory.[59] This was a review of a book on mechanics published in London in 1815, communicated to Silliman for his journal. Overall, the reviewer recommended the book, although he had some very critical comments about Gregory's use of analytic rather than synthetic methods.

The reviewer lamented that in Gregory's demonstration of a fundamental principle of mechanics the "analytical and far-fetched method of the moderns"[60] was used. This method, "however valuable in itself, is certainly defective in producing mental illuminations, or a complete conviction of the truth, and is therefore improper in a work calculated for learners." He went on to say,

> If, after the discovery of a mathematical truth, a demonstration be necessary at all, it is necessary that the reasoning be clear, and evident, at every step; but the analytical process is the very reverse of this, it con-

---

Styles of Algebra in the Early American Republic," in David Rowe and John Mcleary, eds., *History of Modern Mathematics*, vol. 1, (Boston: Academic Press, 1989), pp. 125–154.

[59] 1824, 7:72–85.

[60] *Ibid.*, p. 77.

sisting of mechanical maneuvers of symbols and abstract quantities.

Furthermore, the conclusions of analysis

> ... are drawn from the metaphysical and refined doctrine of ultimate and vanishing quantities, which are considered as difficult of conception even by mathematicians, and wholly unintelligible to learners. Such are all those pretended demonstrations by the differential calculus, generally used by the continental mathematicians of Europe, and now without judgement attempted to be introduced among the English population throughout the world. To us there appears as much sanity in this new fangled mathematics for demonstrations, as in endeavoring to lay the foundation of a structure at its top.[61]

The reviewer went on to admit that analysis had a place in mathematics as a tool for investigation, but not for demonstration or for teaching basic concepts. Later, the reviewer added,

> It would be difficult, indeed, to conceive why Fluxions should be introduced to prove the fundamental truths of any science, when those of its own are the least evident of any of the branches.[62]

By mentioning both differential calculus and fluxions in his diatribe, the reviewer made clear his disdain for the use of calculus in any form for anything but a tool of discovery. Like many mathematicians of the period, the reviewer believed the ancient synthetic

---

[61] *Ibid.*

[62] *Ibid.*

method of Euclid was the only acceptable path to either mathematical demonstration or education.[63]

In his attack on Gregory's use of modern analysis, the anonymous reviewer reiterated many old arguments against calculus, including both Newton's fluxions and the differential calculus of Leibniz and later continental mathematicians. He objected to the logical foundations, or lack thereof, of the calculus. Interestingly, he used the terms "ultimate and vanishing quantities" in this part of his attack on calculus. These terms were the same as those used by Newton in his attempts to establish the method of fluxions on a solid metaphysical base. The reviewer made several positive allusions to Colin Maclaurin, the British mathematician whose work was thought by many to have established the foundations of fluxions on firmer ground. So, even though his major objections are directed towards the continental mathematicians and their "newfangled" techniques, the author of this review obviously did not accept the use of calculus, whether fluxions or differential calculus, for anything but an avenue of discovery for mathematicians. In summary, he seemed resentful that foreign mathematicians were forcing this new mathematics upon the people of America. It is not difficult to imagine that this author, whoever he was, was probably

---

[63]The ideas of our anonymous reviewer were not unusual and probably not even original. In fact, his complaints against analysis echo similar complaints made by the Scottish mathematician John Leslie in 1809:

> The analytical investigations of the Greek geometers are indeed models of simplicity, clearness and unrivalled elegance;...some of the noblest monuments of human genius. It is a matter of deep regret, that Algebra, or the Modern Analysis, from the mechanical facility of its operations, has contributed, especially on the Continent, to vitiate the taste and destroy the proper relish for the strictness and purity so conspicuous in the ancient method of demonstration. The study of geometrical analysis appears admirably fitted to improve the intellect, by training it to habits of precision, arrangement, and close application. If the taste thus acquired be not allowed to obtain undue ascendancy, it may be transferred with eminent utility to Algebra, which, having shot up prematurely, wants reform in almost every department.

Quoted in Craik, "Geometry versus Analysis," p. 139.

a vocal opponent of the move to bring modern analytical techniques to the American colleges.

Other contributors to Silliman's journal found the use of calculus techniques less dangerous than our unnamed reviewer, yet still took issue with the lack of logical foundation for the new methods. Professor Wallace of Columbia, South Carolina, noted in passing

> ... the principle of exhaustions, of indivisibles, of the theory of limits, of prime and ultimate ratios... are still liable to the objections of Berkeley.[64]

In the same volume, Nathaniel Bowditch agreed that these objections remained valid, while pointing out that Wallace had made a crucial mistake himself in introducing a quantity whose value was zero, yet was later treated in division as non-zero. This, Bowditch concluded, was what Berkeley rightfully called the *"ghost of departed quantity."*[65]

Disagreements over issues involving calculus abounded in *The American Journal of Science and Arts.* Several works appeared in the journal that defended Newton and his method of fluxions. Some continued the century-old debate over who should be credited with the original discovery of calculus, while others defended the foundational aspects of calculus, whether fluxions or differentials. In another anonymous review, this time of Newton's *Principia*, the reviewer defended Newton as the true discoverer of calculus and maintained that Newton's ideas on limits and ultimate ratios were sufficient to place calculus on a solid logical foundation.[66]

---

[64]"Professor Wallace, in reply to the Remarks of B. upon his paper on Algebraic Series," 1825, *9*:98–103. The B. in the title refers to Nathaniel Bowditch.

[65]This is a well-known phrase from Bishop Berkeley's *The Analyst*, probably the most famous and widely quoted attack on the foundations of calculus. One of Berkeley's objections to calculus was that infinitely small, or vanishing, quantities were at one turn treated as finite quantities when used in division, and at the next assumed to be zero.

[66]"A Review of the *Principia* of Newton," 1826, *11*:238–245, continued in 1828, *13*:311–321.

A similar article appeared under the name of Proclus, entitled "On the Principles of Motion, and their use in the higher branches of Mathematics."[67] This was a defense of the foundations of fluxions and differentials, but especially the idea of a fluxion as based on the motion of a point. The author called on the names of Robins, Jurin, and Maclaurin[68] to defend Newton's ideas of fluxions, addressing the detractors of Newton somewhat sarcastically:

> Newton and Leibnitz had laid the foundation of this science on a sufficiently tenable ground; but illustration and more practical arguments were necessary for the less penetrating and profound, the disciples of Berkeley and others.[69]

The author also criticized Lagrange[70] for calling into question the rigor of these methods. In the reviewer's own mind, as in the mind of many mathematicians of the time, the question of logical foundations of calculus had already been put to rest.

In the same volume as this defense of Newton's foundations for calculus, Elizur Wright contributed a lengthy, textbook-like presentation of the method of fluxions.[71] Previously published in the *Memoirs of the Connecticut Academy of Arts and Sciences*, this work approached the methods and foundations of calculus from a strictly Newtonian framework. One year later, Wright contributed

---

[67] 1828, *14*:297–302.

[68] These were three of the more prominent supporters of Isaac Newton in the dispute over priority between Newton and Leibniz.

[69] 1828, *14*:298–299.

[70] Joseph Louis Lagrange was a renowned French mathematician who attempted, with incomplete success, to provide a logical (algebraic) foundation for calculus. There is no indication that American mathematicians were yet aware of the work of Augustin Louis Cauchy, another French mathematician who earlier in the decade had successfully set calculus on a firm foundation with his book *Cours d'analyse* (1821).

[71] "A Theory of Fluxions," 1828, *14*:330–350.

another article on fluxions, "A Discourse on the different views that have been taken of the Theory of Fluxions."[72] This time, Wright's attention focused entirely on addressing the foundations of fluxions. He discussed Newton's ideas of ultimate ratios of finite quantities and his theory of limits. Wright's goal in the essay was one of "clearing away the rubbish, and rendering easy and pleasant to the learner, the entrance to this science, by exhibiting a view of its first principles."[73]

The last major article on calculus appearing in Silliman's journal during our time period was entitled "Solution of a Problem in Fluxions," by Theodore Strong.[74] This article is interesting for several reasons. Although Strong used "fluxions" in his title, he used Leibnizian notation and terminology exclusively throughout the article. The problem Strong proposed to solve was this:

> Supposing that a particle of matter, projected from a given point, in a given direction, with a given velocity, is deflected from its rectilinear course into a curve line; It is required to determine the equations of its motion.[75]

Strong used fairly sophisticated mathematics, at least in comparison to other published work in America. He made use of spherical coordinates and parametric equations, as well as many of the techniques found in Laplace's *Mécanique Céleste*.[76] Strong concluded the solution of this problem in the next volume.[77]

---

[72] 1829, *16*:53–59.

[73] *Ibid.*, p. 58.

[74] 1829, *16*:283–287.

[75] *Ibid.*, p. 283.

[76] Interestingly, Strong cites page numbers in the original French version of Laplace's work. Although Nathaniel Bowditch had completed his English translation by this time, it had not yet been published.

[77] 1830, *17*:69–73.

Although the essays appearing in one journal over a period of several decades may not provide enough evidence to draw ultimate conclusions about the state of American mathematics, several inferences may be made from these discussions of calculus. First, the objective of many of the authors was education. An awareness existed in America that original and important research could not occur unless Americans were educated in the new methods of mathematics. This goal of education explains the rhetoric in many of the articles in *The American Journal of Science and Arts*. It was important to choose the correct form of mathematics, whether the argument be synthetic versus analytic style or Newtonian versus Leibnizian calculus. These authors insisted that there was a correct way to learn mathematics.

Secondly, American mathematicians were attempting to deal with the influx of new mathematical ideas from Europe. At the same time as Americans were struggling with these questions, young British mathematicians were setting into motion the processes by which their nation would join the rest of Europe in modern mathematics. America, which until recently had been dependent on England for its intellectual guidance, was itself beginning the process of assimilating the new methods of analysis.

Finally, the very fact that there were a sufficient number of interested parties to encourage open debate in Silliman's journal bode well for American mathematics.[78] It is difficult to determine

---

[78] Debate in the pages of *The American Journal of Science and Arts* was not limited to questions concerning calculus. From 1824 to 1825, Professor Wallace and Nathaniel Bowditch carried on a disagreement over a paper published by Wallace in Volume VII (1824) in which Wallace seemed to claim the discovery of a new infinite series. Bowditch responded in Volume VIII that the series given by Wallace had actually been found first by Euler and could be found in Lacroix's *Algebra*. A back and forth discussion ensued in which Wallace claimed that much of his original submission, including the original title, was omitted by the editors. Wallace maintained that he had given credit to the original discoverers of the series and had not used the word "new" in his paper. Wallace went on to lecture Bowditch that nothing was ever really "new" just as Newton had borrowed from Wallis and others when he discovered calculus. It was the application of the series that Wallace was claiming as important. Bowditch closed the argument (at least in print) by acknowledging that he

whether these defenses of the outdated methods of calculus were representative of American mathematicians as a whole or simply opinions of those who were prone to publish in Silliman's journal. There were certainly Americans who had already enthusiastically adopted the new methods of analysis, including Nathaniel Bowditch, John Farrar, and A. M. Fisher. No matter the sophistication of the arguments, or the antiquity of the mathematics, the interest generated by mathematical questions pointed towards a growing interest in mathematical research in the United States.

This is not an exhaustive list or discussion of all the mathematical articles appearing in *The American Journal of Science and Arts*. The years 1822–1834 were a time when mathematics, both pure and mixed, played a small but steady role in the life of the journal. For some reason, however, the second half of the decade of the 1830's marked an abrupt halt in the publication of mathematical papers in Silliman's journal. From Volume 26 (1834) to Volume 38 (1840), only one mathematical contribution appeared, that by Theodore Strong. Although the journal continued to publish an occasional article on astronomy, the strong current of mathematics was discontinued.

It is not clear why the publication of mathematics suddenly ceased in Silliman's journal. There were no hints in the pages of the journal itself pointing towards a change in editorial policy, and other scientific papers on various topics continued unabated. A possible explanation might be the appearance of *The Mathematical Miscellany*, a journal established in 1836. Although discontinued in 1839, the journal did receive regular contributions from Theodore Strong (previously one of the most prolific mathemati-

---

had not known of the changes made by the editors to Wallace's original communication. However, Bowditch maintained, his reading of the article led him to believe that Wallace was claiming a new discovery, and that "A comparison of the two methods, shows that Euler's demonstration is identical in principle with that published by Professor Wallace, as B[owditch] asserted in his first communication." 1825, *9*:300.

cal contributors to *The American Journal of Science and Arts*) and from Benjamin Peirce and William Lenhart, two young mathematicians who would become distinguished members of the next generation of American mathematics.[79] With the specialization to come to American science later in the century, general topic scientific journals such as the *Transactions of the American Philosophical Society*, the *Memoirs of the American Academy of Arts and Sciences* and *The American Journal of Science and Arts* played an ever-decreasing role in American mathematics. Mathematicians, like other specialists, would eventually realize the need for their own specialized journals.

5.5 Early Mathematical Journals

As the nineteenth century opened, there were no journals in the United States dedicated exclusively to mathematics. By 1840, several such journals had appeared, but all were short-lived. These journals had several things in common. None lasted more than seven years, with most folding after publishing only a few volumes. None of these journals produced the kind of original mathematical research that would make European mathematicians take notice of their American counterparts. Instead, these journals were typically filled with challenge problems written by and for amateur mathematicians, with an occasional expository contribution from a practicing scientist. Only rarely did a piece of mathematics appear in these journals that could even generously be called research. The last thing that these journals had in common is the most important: they represented the first attempts by American mathematicians

---

[79] For the influence of *The Mathematical Miscellany*, see Deborah Kent, "The Mathematical Miscellany and The Cambridge Miscellany of Mathematics: Closely connected attempts to introduce research-level mathematics in America, 1836–1843," *Historia Mathematica*, 2008, *35*: 102–122.

to form one of the crucial components for the development of a true community of professionals.

The remainder of this chapter will consist of an overview of these first American mathematical journals with a critical eye towards their role in encouraging publication from the small cadre of American mathematicians.

The first periodical dedicated exclusively to mathematics in the United States was *The Mathematical Correspondent*, edited by George Baron. Baron was an English immigrant who arrived in the United States just before the turn of the century.[80] Baron taught briefly at West Point before starting his own academy in New York. In 1804, he founded *The Mathematical Correspondent* with the intent

> To inspire youth with the love of mathematical knowledge, by alluring their attentions to the solutions of pleasant and curious questions—and to promote the cultivation of mathematics, by opening a channel for the ready conveyance of discoveries and improvements, from one mathematician to another.[81]

We see, then, two motives common to other scientific journals in the United States: to educate the youth of America and to provide an avenue for the publication of original work.

When Baron arrived in the United States, he was surprised to find a low state of mathematical knowledge as compared to Europe:

> When we consider the great exertions of learned men to disseminate mathematical information in other countries,

---

[80] For biographical information on George Baron, as well as his role in *The Mathematical Correspondent*, see Edward R. Hogan, "George Baron and the *Mathematical Correspondent*," *Historia Mathematica*, 1976, *3*:403–415 and David E. Zitarelli, "The Bicentennial of American Mathematics Journals," *The College Mathematics Journal*, 2005, *36*: 2–15.

[81] *Mathematical Correspondent*, 1804, *1*:title page.

we must be surprised to find that this kind of knowledge is most shamefully neglected in the United States of America.[82]

As a recent immigrant from England, Baron understandably undertook to pattern his new journal after those with which he was familiar, saying: "*The Mathematical Correspondent* will be conducted on the same plan as the European works."[83] Baron's grand plan was to create a European-style mathematical journal to fill the void in mathematical knowledge he found when he came to the United States.

Baron often expressed his displeasure at the level of mathematical learning in the United States. He used *The Mathematical Correspondent* as a platform to lecture on his interpretations of basic mathematical concepts and also to attack the work of other mathematicians. In the opening article of the first *Mathematical Correspondent*, Baron published a lecture on the basic principles of arithmetic. Later, he wrote an exposition on the meaning of "power" in mathematics. Finally, Baron attempted to clarify the definition of proportional.

At first glance, these articles seem to be simply attempts by Baron to educate his American readers. Further reading of Baron's published works, however, reveals a critical and arrogant attitude, much of which appears to be an air of British superiority.[84] Both Hogan and Zitarelli discuss Baron's biting criticisms of Bowditch's *New American Practical Navigator* and Jared Mansfield's *Essays, Mathematical and Physical*, two of America's most important pub-

---

[82] *Ibid.*, p. iii.

[83] *Ibid.*

[84] It is interesting to observe that the only exposition on calculus to appear in the pages of *The Mathematical Correspondent* was an "Essay on Fluxions," by Walter Folger. Folger's essay was taken directly from Vince's *Fluxions*, a popular textbook in Britain.

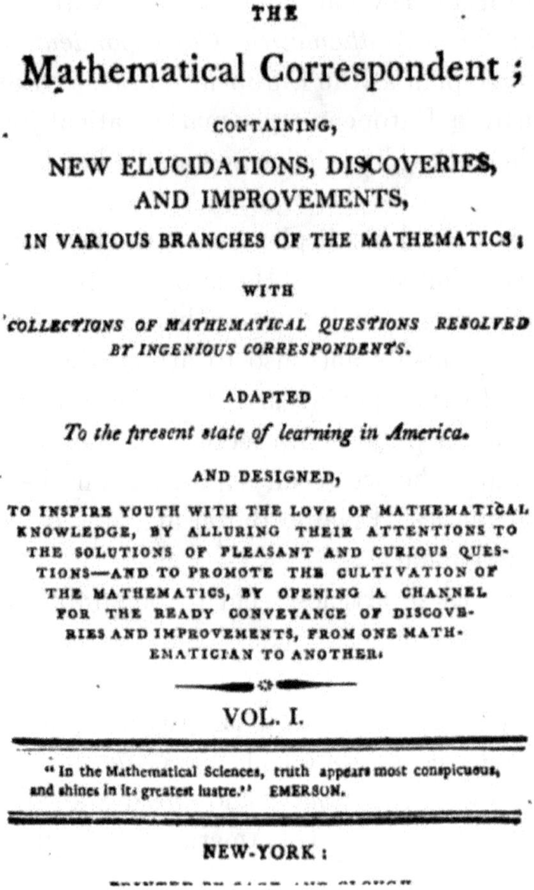

THE
Mathematical Correspondent;

CONTAINING,

NEW ELUCIDATIONS, DISCOVERIES, AND IMPROVEMENTS,

IN VARIOUS BRANCHES OF THE MATHEMATICS;

WITH

COLLECTIONS OF MATHEMATICAL QUESTIONS RESOLVED BY INGENIOUS CORRESPONDENTS.

ADAPTED

To the present state of learning in America.

AND DESIGNED,

TO INSPIRE YOUTH WITH THE LOVE OF MATHEMATICAL KNOWLEDGE, BY ALLURING THEIR ATTENTIONS TO THE SOLUTIONS OF PLEASANT AND CURIOUS QUESTIONS—AND TO PROMOTE THE CULTIVATION OF THE MATHEMATICS, BY OPENING A CHANNEL FOR THE READY CONVEYANCE OF DISCOVERIES AND IMPROVEMENTS, FROM ONE MATHEMATICIAN TO ANOTHER.

VOL. I.

"In the Mathematical Sciences, truth appears most conspicuous, and shines in its greatest lustre." EMERSON.

NEW-YORK:

Figure 5.5.1: Title page of the first edition of *The Mathematical Correspondent* (1804)

lications in the early years of the century.[85] More evidence that Baron's scorn for these works might be a case of nationalism is that after deriding Bowditch's monumental work on navigation, he proceeded to recommend an alternative work on navigation by his fellow countryman, Andrew Mackay.[86]

In addition to Baron's criticisms, an anonymous author, A. Rabbit, contributed several articles critical of American arithmetic texts. In one such attack, Rabbit stated:

> In this country authors of arithmetic have lately sprung up like a parcel of mushrooms, and it would have been well for the young and rising generation had the former been as harmless as the latter. These upstart authors have most perniciously corrupted, distorted, and degraded the noble and useful science of numbers, and metamorphosed our sons into mere counting machines moving according to a heterogeneous collection of unscientific and stupid rules. A good book in arithmetic is much wanted in America, but so long as the wretched productions of Pike, Walsh, Shepherd, and Co. are encouraged, we cannot expect a man of talents to enroll his name in our list of numerical authors.[87]

Pike's *Arithmetic*, in particular, was a popular work in America and an attack against it could be seen as an attack against American abilities and therefore would cut directly to national pride.

Baron's caustic pen and seemingly anti-American attitude may have contributed to his resignation as editor of *The Mathematical Correspondent*. Whatever the reason, Baron announced after only

---

[85] Hogan, "George Baron," p. 406. Zitarelli, "Bicentennial," p. 6.

[86] *Ibid.*, p. 409.

[87] Quoted in *Ibid.*, pp. 406–407.

one volume that he could not edit future numbers of the journal due to other engagements. He left the journal in the hands of Robert Adrain, who was to become over the next several decades a central figure in the American efforts to sustain a specialized mathematical journal.

In his preface to the second volume (his first as editor) of *The Mathematical Correspondent*, Adrain apologized to the writers who had been offended by Baron and assured his readership that these types of personal attacks would no longer occur in the pages of the *Correspondent*. He also defended the format of the journal, maintaining that proposing and solving problems in a public forum had served well for many famous mathematicians, including the Bernoullis, Huygens, Newton, Euler, Lagrange, and others. Adrain understood, much better than had Baron, that bickering and personal vendettas would not advance the cause of mathematics in America.

The problems proposed and the articles published in *The Mathematical Correspondent* were for the most part simple and uninteresting from a mathematical point of view. In addition to Baron's essay on the principles of arithmetic, the first installation of the new journal contained two worked problems, one a simple ratio problem and the other a geometric proposition, as well as problems posed to the readers from various people. The answers to previous problems, as well as new challenge problems, made up the bulk of future issues of *The Mathematical Correspondent*. The most significant problem appearing in the pages of the journal was one proposed and solved by Adrain involving elliptic integrals in the first volume of the *Correspondent*. The problem was evidently forgotten until 1860 when it was independently solved by Alfred Clebsch.[88] In addition, it was in the pages of *The Mathematical*

---

[88] Julian L. Coolidge, "Robert Adrain, and the Beginnings of American Mathematics," *American Mathematical Monthly*, 1926, *33*:61–76, on p. 65.

Correspondent that Adrain introduced Americans to Diophantine Algebra with the first article to appear on the subject in the United States.[89]

The Mathematical Correspondent was, like many journals of the day both in the United States and abroad, short-lived. After Robert Adrain became editor in 1807, he was only able to maintain circulation for one additional volume. The importance of *The Mathematical Correspondent*, then, does not rest on its longevity or on its contents. It is important, however, as a symbol. *The Mathematical Correspondent* represents a first effort by American (albeit emigrant) mathematicians to organize themselves into a self-contained group independent of the larger scientific population.

Although the lack of a critical mass of mathematicians was almost certainly the major factor in the demise of *The Mathematical Correspondent*, the fact that there were nearly 350 subscribers (although not all of them were paying subscribers) must have been encouraging for Adrain. Participation in the problem-solving section of the journal was also fairly consistent. On an average, approximately twenty people submitted solutions to problems in each issue. Although most of these contributions were relatively unsophisticated, future leaders of American mathematics such as William Lenhart, Robert Patterson, Jr., and Adrain himself contributed solutions. Adrain's participation as a contributor and later as an editor of *The Mathematical Correspondent* may have whet his appetite for such work, as he spent much of the next three decades in various attempts to establish mathematical journals.

Born in Ireland in 1775, Robert Adrain escaped to the United States after the Irish rebellion of 1798.[90] Adrain taught at various

---

[89] *Ibid.*, p. 66.

[90] Biographical information on Robert Adrain comes primarily from D.J. Struik, "Robert Adrain," in Gillispie, *Dictionary*; and Greene, *American Science*, pp. 132–134. A contemporary account of Adrain's life is in his obituary in the *U.S. Magazine and Democratic Review*, 1844, *14*:646–652. Summaries of Adrain's work and his contributions to American mathe-

institutions in America, including Queen's College (now Rutgers), Columbia College, and the University of Pennsylvania, where he also served as vice-provost from 1828–1834. As noted, Adrain assumed the editorial duties for *The Mathematical Correspondent* after those duties were relinquished by George Baron in 1807. After *The Mathematical Correspondent* folded, Adrain established several other journals, including *The Analyst, or Mathematical Museum* (1808), which he attempted to revive as simply *The Analyst* in 1814. Adrain's efforts to provide an outlet for research and at the same time bring interesting problems to Americans interested in mathematics, continued as he edited the first six volumes of *The Mathematical Diary*, a journal that survived from 1825 to 1833.

In addition to his work in editing journals, Adrain published many articles in the pages of his journals and in other periodicals.[91] He also prepared the American edition of Hutton's *Course of Mathematics* in 1812, a very popular textbook in American colleges for many years.[92] Adrain's work on Hutton's text evolved into a large number of explanatory notes as well as several additions and changes to Hutton's original definitions.

---

matics are in Coolidge, "Robert Adrain," and Edward R. Hogan, "Robert Adrain: American Mathematician," *Historia Mathematica*, 1977, *4*:157–172.

[91] These include *The Portico, The Scientific Journal* (using the pseudonym "Analyticus"), the *Transactions of the American Philosophical Society, The Ladies' and Gentlemen's Diary or U.S. Almanac* and the problem sections of the *New York Mirror and Ladies' Literary Gazette*. Hogan, "Robert Adrain," p. 159.

[92] Charles Hutton was a British professor of mathematics at the Royal Military Academy for 34 years. He prepared a work entitled *A Course of Mathematics for the Cadets of the Royal Military Academy* between 1798 and 1801. *Course of Mathematics* was a text that covered the rudiments of most of the mathematical subjects deemed important for a sound education, including arithmetic, the use of logarithms, algebra, geometry, plane trigonometry and calculus (the doctrine of fluxions). Its wide coverage of mathematics and its concise style did cause many an American student, like their British counterparts before them, to flinch at the thought of mastering such a book. Coolidge dryly notes "One can not wonder at the bravery of the British officers at Waterloo if they had mastered Hutton." Coolidge, "Robert Adrain," p. 70.

Figure 5.5.2: Robert Adrain

In the third American edition of Hutton's *Course of Mathematics*, Adrain added a section on descriptive geometry, a relatively new development in mathematics.[93] The section on calculus relied entirely on Newtonian notation and the method of fluxions. Even the fifth edition, revised by Olinthus Gregory and published in 1831, continued to use Newton's notation.

Robert Adrain and Nathaniel Bowditch have been called the first creative mathematicians in America. Adrain's talents allowed him to rise quickly in the small circle of American mathematicians. In addition to his appointments at Rutgers, Columbia, and Pennsylvania, Adrain was elected to the American Philosophical Society in 1812 and to the American Academy of Arts and Sciences in 1813.

Adrain was not known as a successful teacher. He encountered riotous students at Pennsylvania, possibly due to his own classroom style. A former student of Adrain's wrote that although Adrain was "Eminent in his chosen line of study, and very fond of its pursuit, he had little or no faculty of imparting his knowledge to others." The former student, Dr. Benjamin Haight, continued:

> If one was thoroughly prepared in his recitation, all was well, but if the student was in doubt, or needed a word of explanation in a difficult problem, he not only did not get assistance [from Adrain] but was send [sic] down with some remark of the sort "If you cannot understand Euclid, Dearie (a term he frequently used when out of temper) I cannot explain it to you." The consequence was that a small portion of the class only could keep up with his course, those who had entered college thoroughly versed in the elements of mathematics, and who studied very

---

[93] Coolidge compares Adrain's section on Descriptive Geometry to the works of Monge, the originator of the subject, and to Lacroix and Crozet (who published his own text on descriptive geometry at West Point a year earlier). Coolidge finds the most similarities between Adrain and Crozet, although he finds enough differences to credit Adrain with "[thinking] through the whole subject in his own way." *Ibid.*, p. 71.

diligently after they had entered his lecture room, in my class not more than one fifth of the number. I ought to add, however, that those who went to him in private always found him kind in manner and ready to answer their questions and help them out of their difficulties.[94]

We see in this picture of Adrain a man dedicated to his mathematics with little patience for any student less dedicated than himself.

After the demise of *The Mathematical Correspondent*, Adrain immediately plunged back into his work by launching *The Analyst, or Mathematical Museum* in 1808. In fact, the first number of *The Analyst, or Mathematical Museum* was identical to the last number of *The Mathematical Correspondent*, with a few corrections made to the earlier publication. Adrain began with the same preface as he wrote for the second volume of *The Mathematical Correspondent* after he had replaced Baron as the editor of that journal. In his defense of the format of the journal (proposed problems and their solutions), Adrain maintained that the prize questions should "be useful in improving some important theory little known, or in discovering some new one, or, lastly, in giving some rules of practical application."[95]

Interestingly, Adrain's emphasis on application appeared to be much less than that of other journals, especially those non-specialized journals previously discussed. Adrain specifically listed application as a last, rather than first, option for publication of prize problems. At the very least, it appears that Robert Adrain placed a value on pure mathematics equal to that of mixed, or applied, mathematics. This seemingly inconsequential remark in the preface of Adrain's new journal does indicate that at least he,

---

[94] Quoted in Hogan, "Robert Adrain," pp. 163–164.

[95] *The Analyst, or Mathematical Museum*, 1808, *1*:v.

if not others, was interested in supporting mathematicians doing research in pure mathematics.[96]

Although extremely short-lived, Adrain's 1808 periodical was an improvement upon Baron's journal from only a few years previous. Unlike Baron, Adrain did not use his journal as a personal platform for attacking other mathematicians and pontificating on his ideas of what mathematics should be. Instead, Adrain made the first attempt to bring a professional mathematics journal to America.[97] *The Analyst, or Mathematical Museum* contained a lively exchange of questions, solutions, and offers of prizes from notable mathematicians such as John Gummere, Robert Patterson, Ferdinand Hassler, Nathaniel Bowditch,[98] and Adrain himself. This trend, that given any opportunity American mathematicians would support new or existing journals with their contributions, would continue through a series of failed journals. Unfortunately, this handful of supporters would prove insufficient for a mathematical journal to survive.

The most remarkable work to appear in *The Analyst, or Mathematical Museum*, and possibly the most remarkable original mathematical work from an American in the first half of the nineteenth century, was Adrain's solution of the error distribution problem found in the fourth number of his journal. The problem, often called "combination of observations" in the nineteenth century, was motivated by the need to find a way to combine numerous obser-

---

[96] This conclusion conflicts with Hogan's position that Adrain "considered applications to be of the greatest importance," even quoting a letter written by Adrian to that effect. I believe that Adrain's work shows a man who was passionate about pure mathematics, but realistic in his understanding of what sort of mathematics would interest the American mathematical community.

[97] In fact, Baron's *Mathematical Correspondent* and Adrain's *Analyst* both predate the *American Mineralogical Journal* (founded in 1810), which Greene calls the "first specialized scientific publication in the United States apart from medical journals." Greene, *American Science*, p. 91.

[98] Bowditch submitted solutions to all ten problems presented in the second number of *The Analyst, or Mathematical Museum*, including the answer to the prize problem.

vations in fields such as astronomy and geodesy into one result. In the first decade of the nineteenth century the work of two men, Adrien-Marie Legendre and Carl Friedrich Gauss, represented the most important advance in the field of the statistical combination of observations.

In 1805, Legendre published a treatise on comets, *Nouvelles méthodes pour la détermination des orbites des comètes.* In a nine-page appendix to his work, Legendre published the first discussion of the method of least squares. Essentially, this method uses calculus to minimize the sum of the squares of the "errors," or the difference between the observed values and the assumed "best" solution. This method, which remains the method of choice for such problems today, has been called "the dominant theme—the leitmotif—of nineteenth-century statistics."[99]

Four years later, in 1809, Gauss published a work entitled *Theoria Motus Corporum Coelestium in Sectionibus Conicis Solum Ambientium,* in which he claimed he had been using the method of least squares since 1795.[100] In this work, Gauss developed a more acceptable theoretical basis for the method of least squares than had Legendre.[101] We see, then, the problem of distribution of errors in observations was very much in question when Adrain first addressed it in 1808.

Adrain's solution was motivated by a problem submitted by Robert Patterson in the second number of *The Analyst, or Mathematical Museum.* Patterson's problem involved the errors produced when measuring the sides and angles of a polygon. Adrain's

---

[99]Stephen Stigler, *The History of Statistics* (Cambridge: Cambridge University Press, 1990), p.11.

[100]A prolonged priority dispute ensued between Legendre and Gauss.

[101]Stigler points out that, from a modern point of view, Gauss's derivation, although better than Legendre's, still contains circular arguments. Gauss's work on the subject did not close the question. Stigler, *History*, p. 141.

solution to this problem was partially in response to a solution submitted by Nathaniel Bowditch, which Adrain found lacking in generality. Although Adrain's solution was also lacking in some of its basic assumptions, it did represent the first solution to the problem to appear in print, preceding Gauss (after whom the solution came to be named) by a year. It is unclear whether Adrain knew of Gauss' work on the normal distribution of errors, or if he was familiar with Legendre's method of least squares. Stephen Stigler believes that it was not an independent discovery on the part of Adrain.[102]

Following his derivation of the error "law," Adrain derived the method of least squares, primarily for use in solving Patterson's polygon problem. At some point, Adrain must have learned of Legendre's derivation of the method of least squares, as Legendre's work was a part of Adrain's library.[103] It is not clear, however, when Adrain first read Legendre's work. Regardless of the timing of his discovery, Adrain's use of his version of the error law to deduce the method of least squares represents a highpoint in American mathematics in the first half of the nineteenth century. This and other original work are more impressive when it is remembered that Adrain did not have many of the advantages of the typical European mathematician. His limited library and contact with European mathematicians, and his constant struggle to find suitable employment that would provide the leisure time needed to sustain his research, all make Adrain's accomplishments more remarkable.[104]

---

[102] Stephen Stigler, "Mathematical Statistics in the Early States," in Peter Duren, ed., *A Century of Mathematics in America*, Volume 3 (Providence, RI: American Mathematical Society, 1988), pp. 537–564, on p. 542.

[103] Coolidge, "Robert Adrain," p. 69.

[104] Tocqueville wrote, "The higher sciences or the higher parts of all sciences require meditation above everything else. But nothing is less conducive to meditation than the setup of a democratic society." Tocqueville, *Democracy*, p. 460. This was certainly true for Adrain,

Adrain's publication of the solution to the error problem[105] went unnoticed by other mathematicians. Evidently, American mathematicians did not recognize the importance of his work, and European mathematicians may have never even seen it. No mention of Adrain's proof may be found until Cleveland Abbe rediscovered his work and published it in *The American Journal of Science* some 63 years later.[106] It should be noted that although Adrain's proof of "Gauss' law" was flawed, other subsequent proofs, although more acceptable to mathematicians than Adrain's, have also fallen short of full acceptability.

Both *The Mathematical Correspondent* and *The Analyst, or Mathematical Museum* are of historical significance because they represent the first attempts to establish a mathematical journal in the United States. Many more mathematical journals would be born and would die a premature death before success would be achieved in establishing a long-lasting journal dedicated to American mathematics. A few other journals, each established only to vanish before 1840, will now be briefly mentioned.

Robert Adrain's next attempt to begin a sustainable mathematical journal came in 1814 when he founded *The Analyst*. This journal had the shortest run of any of the journals with which Adrain was involved, publishing only one number. Adrain expressed similar reasons for beginning this journal as he, and Baron before him, had expressed in previous journal start-ups:

> In a word, the objects of the work are to accelerate the progress of the young, excited to action the power of

---

as it had been for Nathaniel Bowditch. The time required for meditation was exceedingly difficult to find for men struggling to make a living in America.

[105] Actually, Adrain gave two proofs. It is the first that is usually considered the stronger of the two.

[106] *Journal*, 1871, *1*(3rd series):41.

> genius, display in a proper light those talents, which should be known to the public, and afford a place for such improvements and discoveries as may be useful to mankind.[107]

Adrain, like other teachers and would-be researchers of his time, saw a profound need not only to educate Americans in his subject, but also to provide a place where original research might be published in the United States.

The Mathematical Diary, also edited by Adrain,[108] made its appearance in 1825. Although it lasted for eight years, much longer than previous mathematics journals, it catered "to the tastes of problem solvers, generally of the lower grade, and exercis[ed] no particular influence upon the mathematics of the country."[109] Hogan agrees with this assessment, postulating that

> ...it seems reasonable that Adrain was attempting to seek a wider audience for *The Mathematical Diary* than was enjoyed by *The Analyst*, at the cost of a lower over-all quality to the journal.[110]

In spite of this negative judgment of the value of *The Mathematical Diary*, it did contain a few interesting problems allowing men such as Adrain, Strong, and Bowditch to exhibit their grasp of some difficult mathematical subjects, such as the calculus of variations. It was also in this journal that a young mathematician by the name of Benjamin Peirce published his first mathematical work.[111]

---

[107] Robert Adrain, *The Analyst*, 1814, *1*:iv.

[108] Adrain edited the first six volumes, with James Ryan succeeding him as editor for the remaining volumes.

[109] David Eugene Smith, "Early American Mathematical Periodicals," *Scripta Mathematica*, 1932–1933, *1*:277–286, on p. 281.

[110] Hogan, "Robert Adrain," p. 160.

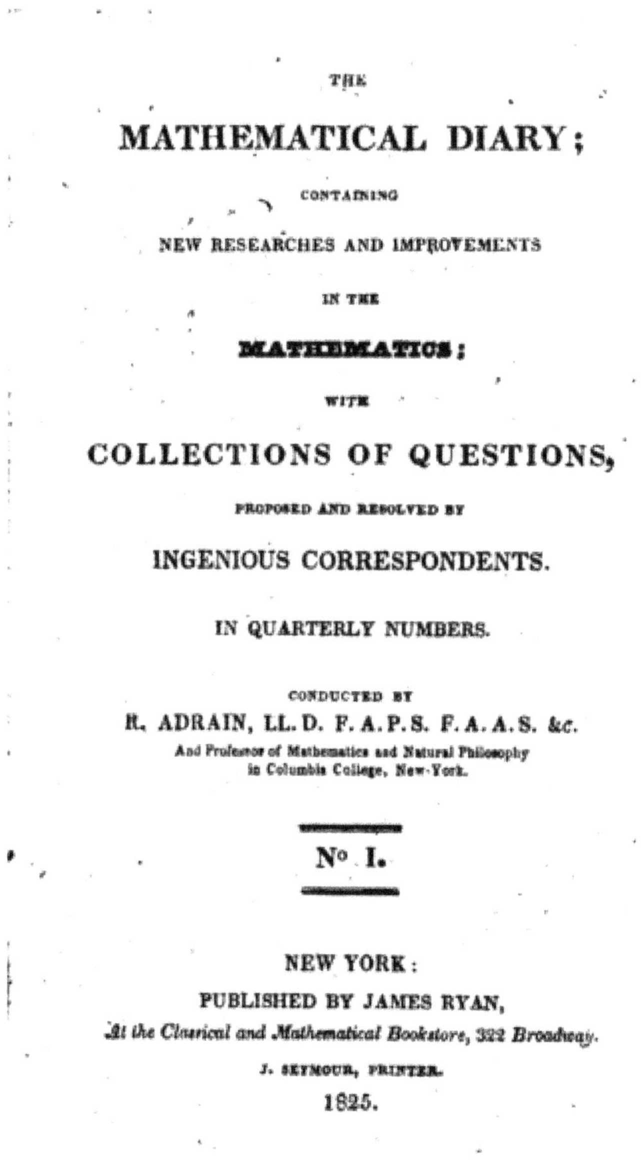

THE MATHEMATICAL DIARY;

CONTAINING

NEW RESEARCHES AND IMPROVEMENTS

IN THE

MATHEMATICS;

WITH

COLLECTIONS OF QUESTIONS,

PROPOSED AND RESOLVED BY

INGENIOUS CORRESPONDENTS.

IN QUARTERLY NUMBERS.

CONDUCTED BY

R. ADRAIN, LL. D. F. A. P. S. F. A. A. S. &c.

And Professor of Mathematics and Natural Philosophy
in Columbia College, New-York.

No. I.

NEW YORK:
PUBLISHED BY JAMES RYAN,
At the Classical and Mathematical Bookstore, 322 Broadway.

J. SEYMOUR, PRINTER.

1825.

Figure 5.5.3: Title page of the first edition of *The Mathematical Diary* (1825)

The Mathematical Miscellany, edited by Charles Gill, was published semi-annually from 1836 to 1839. Gill, like Baron and Adrain before him, was an immigrant, coming to the United States from England in 1830. It seems that native-born mathematicians who might have been in positions to facilitate the publication of mathematical journals did not have the background experiences of Gill, Adrain, and Baron. Although British mathematics continued to struggle in its transition to modern continental methods, immigrants from Britain were accustomed to having outlets for their mathematical work and sought to pattern journals they founded in the United States after familiar periodicals. For example, both Baron and Adrain referred to *The Ladies' Diary* as inspirations for their journals, while Gill contributed to *The Ladies' Diary* while still living in England.[112] It was against this background that immigrants led the way in the mathematical journals that appeared in the United States in the first part of the nineteenth century.

The Mathematical Miscellany was much like its predecessors in many ways. It was primarily a journal for printing the solutions to problems, although it did publish occasional articles from American mathematicians as well as reprints of European publications. *The Mathematical Miscellany*, like previous mathematical journals, was supported by the best American mathematicians of the day. Whereas *The Mathematical Correspondent* and the several journals edited by Robert Adrain counted among their contributors Nathaniel Bowditch, Robert Patterson, Ferdinand Hassler, and Adrain, *The Mathematical Miscellany* included contributions from the next generation of mathematicians like Benjamin Peirce, Theodore Strong, and William Lenhart. Because this new genera-

---

[111] Stigler, *Mathematical Statistics*, p. 542.

[112] Edward R. Hogan, "The Mathematical Miscellany (1836–1839)," *Historia Mathematica*, 1985, *12*:245–257, on p. 246. See also Kent, "The Mathematical Miscellany," 106–111.

tion of mathematicians had naturally grown in their mathematical sophistication, so *The Mathematical Miscellany* had grown in its sophistication in comparison to its predecessors.

The demise of *The Mathematical Miscellany* may be attributed, at least partly, to the same reason as the demise of its predecessors. There were simply not enough interested parties in the United States to keep such an endeavor going. Hogan estimates that the circulation of the *Miscellany* never exceeded 100,[113] in spite of Gill's attempts to include less advanced readers with a junior problem section. In addition to the paltry number of subscribers, a large portion of the contributors to the *Miscellany* came from the immediate area in New York in which Gill resided, making the *Miscellany* in many ways a regional, rather than a national, undertaking. This condition was further exacerbated by the lack of participation from college professors around the country. Despite the support of Peirce at Harvard and Strong at Rutgers, few men who held professorships of mathematics were interested in or capable of original research or even problem solving.

If the success of American mathematicians in establishing a repository for their research is based upon the permanence of the journals or the importance of the work appearing therein, then we must judge their efforts to be a failure. However, few enterprises succeed without failed attempts on which to base future efforts. The existence of science journals such as the *Transactions of the American Philosophical Society*, the *Memoirs of the American Academy of Arts and Science,* and *The American Journal of Science and Arts*, provided at the very least a stopgap solution to the problem of publishing original mathematical work.

The attempts by immigrants such as George Baron, Robert Adrain, and Charles Gill to establish a permanent mathematical journal gave their American-born successors a model to emulate as

---

[113] *Ibid.*, p. 251.

they continued throughout the century to experiment with various formats in new mathematical journals. These pioneers paved the way for the eventual establishment of permanent modes of publication for American mathematics.

Chapter 6

The French Connection: Building a Foundation on Modern Mathematics

Previous chapters have addressed in some detail the failures and successes of the Bowditch generation with regards to the first three criteria for the development of a research community—a critical mass of practitioners, the ability to exchange ideas, and availability of financial support for researchers. This chapter will concentrate on the fourth criterion, the educational opportunities available. In particular, we will investigate the first attempts to bring modern mathematical techniques to the American college through the introduction of translations of French textbooks.

As a colony of Great Britain until the last quarter of the eighteenth century, America naturally looked to the homeland for its intellectual and cultural leadership. The ties to Britain were forged through common language, religion, and social and political ideologies. The ties that bound America to Britain in science and mathematics lasted well into the first decades of the nineteenth century.

Sometime after the turn of the century, however, American mathematicians began to discover, and then slowly embrace, the mathematics of continental Europe, particularly the great advances of the French. Helena Pycior has argued that, while pre-Revolutionary America was homogeneous in its reliance on British mathematical texts, post-war America was a heterogeneous mix in which "American educators pursued the active roles of judging, adapting and synthesizing the British and the French mathematics, especially the rival synthetic and analytic styles of algebra."[1] The emergence of this phenomenon laid the groundwork for the modernization of American mathematics and made possible the emergence of American mathematicians who were equipped to compete with their European counterparts on a level playing field.

6.1 French Influence on American Science

In spite of its ties to England, eighteenth-century America was also influenced by French culture. The American Academy of Arts and Sciences, patterned after the Royal Society of London, was named, however, to reflect honor on the Paris Academy of Sciences.[2] Even the Academy's choice for the name of its official publication, the *Memoirs*, carries a French connotation.[3]

The French influence, if it were not for a set of unfortunate circumstances, might have been even greater. Beginning in 1788, a serious attempt was made to found an institution in the United States

---

[1] Helena M Pycior, "British Synthetic vs. French Analytic Styles of Algebra in the Early American Republic," in David E. Rowe and John McCleary, eds., *The History of Modern Mathematics*, vol. 1, (Boston: Academic Press, 1989), pp. 125–154, on p. 126.

[2] Brooke Hindle, *The Pursuit of Science in Revolutionary America* (Chapel Hill: University of North Carolina Press, 1956), p. 264.

[3] The Academy's chief rival in America, the American Philosophical Society, had followed the lead of the Royal Society of London in naming its publication the *Transactions*.

that would mimic the Academy in Paris. L'Académie des États-Unis de l'Amérique was to be centered at Richmond, Virginia, with satellite affiliates in Baltimore, Philadelphia, and New York.[4] The society was the brainchild of John Page, Lieutenant Governor of Virginia, and Chevalier Quesnay de Beaurepaire, a Frenchman who served in the American Revolution. Chevalier Quesnay's plans included a large organization of French scientists and artists who would be imported into the United States to staff the Academy. In spite of the fact that a significant amount of money was raised and a building was constructed to house the Academy, the project failed before it ever really started. Several factors, primarily the beginning of the French Revolution, conspired to doom the American version of the Paris Academy.[5]

Interest in French culture and French science soared in America during the later part of the eighteenth and early part of the nineteenth centuries. Already appreciative of the help from the French during the American Revolution, Americans followed with interest—and often with admiration—the French Revolution and the subsequent reorganization of higher education in France. Thomas Jefferson was symbolic of the esteem many Americans held for the French.[6] As early as 1799, Jefferson, then governor of Virginia, was instrumental in the introduction of the study of French into the curriculum at William and Mary, his alma mater.[7]

---

[4] Ralph Bates, *Scientific Societies in the United States* (Cambridge, MA: The MIT Press, 1965), p. 11.

[5] *Ibid.*, pp. 12–13.

[6] Of course this esteem was not universal. Many of Jefferson's Federalist opponents were wary, and sometimes hostile, to Jefferson's dealings with the French. See Lawrence S. Kaplan, *Jefferson and France: An Essay on Politics and Political Ideas* (New Haven: Yale University Press, 1967); Alf J. Mapp, Jr., *Thomas Jefferson: Passionate Pilgrim* (New York: Madison Books, 1991); and Donald R. Hickey, *The War of 1812* (Urbana and Chicago: University of Illinois Press, 1989).

[7] Kaplan, *Jefferson and France*, p. 15. Many years later, Jefferson showed that he was still convinced of the importance of French for a modern education. In 1819, Jefferson wrote

Sometimes Jefferson's identification with the French bordered on Anglophobia.[8] His biographers agree, however, that as Ambassador to France, Secretary of State, Vice-President, and President, Jefferson always attempted to work both Britain and France to the advantage of the young republic. In fact, in a statement contrary to the public's perception of Jefferson as a Francophile, Jefferson told David Erskine, British Minister to the United States, "Americans really did not prefer the French to their brothers in blood, culture and language [the English]."[9] This dichotomy in the political feelings towards the French and the English was also felt in the realm of science.

American men of science were appreciative of—and sometimes a little jealous of—the opportunities the reforms of the Revolution afforded French scientists.[10] Jasper Adams, a professor of mathematics and natural philosophy at Brown from 1819–1824, maintained that in spite of the fact that the French philosophers and mathematicians were forced to concentrate on science which might prove useful to the people of France, they were benefited greatly by

> ...the establishment of the National Institute [which] concentrated the talents of the nation, and the pensions

---

to Madison that in preparation for the opening of the University of Virginia, he had invited a French teacher by the name of Mr. Stack to Charlottesville to set up a school to prepare future students for the University. In addition, Jefferson convinced a man by the name of LaPorte to open a boarding house for the students in which only French would be spoken. James Morton Smith, ed., *The Republic of Letters: The Correspondence between Thomas Jefferson and James Madison 1776–1826*, pp. 1810–1813.

[8] Hickey finds that this Anglophobia "originally kindled by the American Revolution, was further influenced by the War of 1812." Hickey, *War*, p. 305.

[9] Mapp, *Thomas Jefferson*, p. 175.

[10] Of course this jealousy did not apply to the plight of the many French scientists persecuted, imprisoned, and even executed during the French Revolution and the subsequent Reign of Terror. These included Lazare Carnot (forced into exile), the Marquis de Condorcet (died in prison), and Antoine Lavoisier (guillotined), among others.

and high honors which were liberally bestowed, especially upon those who successfully cultivated the exact sciences.[11]

This turn of events "gave an astonishing impulse to mathematical learning [in France]."[12] Adams went on to credit these circumstances with producing the important mathematical and scientific works of Lacroix, Legendre, Laplace, Lagrange, Poisson, and many other French scientists. He concluded by stating:

> The exact sciences are vastly indebted to the French revolution and its long train of consequences, whatever may be its ultimate effect upon the progress of knowledge in general. The science of calculation is now investigated with such resources, that almost nothing is too complicated, or too stubborn to yield to its power.[13]

Sometimes American mathematicians appeared envious of other scientists practicing their art in societies more supportive of such endeavors.

6.2 British versus French: Conflicting Mathematical Styles

It should come as no surprise that American mathematicians in the first part of the nineteenth century began to turn to the French style

---

[11] From an anonymous review of John Farrar's translation of Legendre's *Elements of Geometry*, in the *American Journal of Science and Arts*, 1823, *6*:283–301, on pp. 284–285. Amy Ackerberg identifies the author as Adams based on a letter from John Farrar to Josiah Quincy. Ackerberg, *Mathematics is a Gentleman's Art: Analysis and Synthesis in American College Geometry Teaching, 1790–1840*, Ph.D. dissertation (Iowa State University, 2000).

[12] Adams, *Journal*, 1823, *6*:285.

[13] *Ibid.*

of mathematics.[14] It was about the same time that British mathematicians began to emerge from their century-old, self-imposed exile from the mainstream of European mathematics. The roots of the British adoption of the new mathematics are usually traced to the founding of the Analytical Society in Cambridge in 1812 by a group of University students led by Charles Babbage, John Herschel, and George Peacock. This group set out to do more than simply change from Newtonian notation to Leibnizian; they were intent on changing the whole foundation upon which mathematical education in England was based.[15] In particular, the Analytical Society collectively believed that Newton's fluxions should not be considered a part of analysis because it was based on the concept of motion, which is not an algebraic concept.[16] The Analytical Society met with only limited success, and the society quietly disappeared after only two years. However, Herschel, Peacock, and Babbage, together and separately, influenced British science and mathematics for many years.[17]

The three founders of the Analytical Society jointly translated Lacroix's *Traité élémentaire de calcul différentiel et de calcul inté-*

---

[14]Sporadic interest in French mathematics may be traced back into the eighteenth century. For example, in 1788 Walter Minto, a University of Pisa-trained professor at Princeton, translated Bossut's *Cours de mathématiques*. The work was never published and it is not clear how much, if any, the text was used by students at Princeton. David Eugene Smith and Jekuthiel Ginsburg, *A History of Mathematics in American before 1900* (New York: Arno Press, 1980), p. 30.

[15]See, for instance, Susan Faye Cannon, *Science in Culture: The Early Victorian Period* (New York: Dawson and Science History Publications, 1978). Cannon places the work of the Analytic Society within a larger context of what she calls the "Cambridge Network," an informal group of Cambridge students and professors who sought to transform the college into a scholarly institution.

[16]Philip C. Enros, "The Analytical Society (1812–1813): Precursor of the Renewal of Cambridge Mathematics," *Historia Mathematica*, 1983, *10*:24–47, on p. 28.

[17]For a different view of the role that analysis vs. synthesis played in British mathematics, see Amy Ackerberg-Hastings, "John Playfair on British decline in mathematics," *BSHM Bulletin: Journal for the British Society for the History of Mathematics*, 2008, *23*: 81–95.

gral, which became popular with many students at Cambridge.[18] Peacock was appointed moderator of the Senate House examination in 1817 where he introduced questions using the notation of differential calculus.[19] Babbage was awarded the Lucasian chair at Cambridge in 1828. Although Babbage did not contribute significantly to the University in his new position, Wilkes considers his appointment to be an indication "that the reforming party had now won their battle."[20] With Britain beginning the long process to replace its mathematics curriculum with the newer French model, it is little wonder that America was also beginning to also test the waters of change.

Much of the tension in mathematics during the latter part of the eighteenth century and the early decades of the nineteenth century may be traced to the often-conflicting styles of analytic and synthetic mathematics.[21] Two examples of the scholarship relating to the question of synthetics versus analytics serve to give a flavor of the conflict in Britain. Pycior uses the treatment of negative numbers as a means of contrasting the two styles of mathematics. In the British synthetic style negative numbers were "defined," providing first principles on which to build a theory. In the French analytic style, examples were used to build towards an explanation of negative numbers.[22]

Enros presents a social implication for the synthetic/analytic debate. He argues that synthetics was linked to a liberal education in

---

[18] M.V. Wilkes, "Herschel, Peacock, Babbage and the Development of the Cambridge Curriculum," *Notes and Records of the Royal Society of London,* 1990, *44*:205–219, on p. 207.

[19] *Ibid.*, p. 212.

[20] *Ibid.*

[21] See the discussion on this subject in the section "The Calculus Question: Differentials or Fluxions, Synthesis or Analysis" in Chapter 5.

[22] Pycior, "British Synthetic," p. 126.

early nineteenth-century Britain, and thus to the education of gentlemen without any intention of pursuing a career in mathematics. On the other hand, analytics was linked to research, and therefore to professionalism. Enros maintains that the Analytic Society died shortly after it was founded primarily because there were few men in Britain interested in mathematics as a profession.[23] In this sense, we can see that British mathematics was not greatly superior to American mathematics. The British, like the Americans, were struggling to reach a critical mass of interested parties on which to build a mathematical community.

A brief look at Robert Adrain's conflicting attitudes towards French mathematics may shed light on the confusing times in which English-speaking mathematicians found themselves. Hogan presents evidence that Adrain recognized the superiority of French texts such as Lacroix's *Course of Mathematics* and also understood the importance of French mathematicians such as Laplace and Lagrange.[24] In fact, Adrain wrote a manuscript while he was at Columbia College (1813–1826) on differential and integral calculus.[25] In spite of this respect for French mathematics, Adrain edited several British texts, most notably the ever-popular Hutton's *Course of Mathematics*. Hogan conjectures that Adrain preferred to edit English-language works for economic rather than mathematical reasons. It is probably an overstatement to say, as Hogan has, that French mathematics had a "strong influence" on Adrain,[26] while at the same time attributing to Adrain an anony-

---

[23] Enros, "Analytical Society," pp. 41–42.

[24] Edward R. Hogan, "Robert Adrain: American Mathematician," *Historia Mathematica*, 1977, *4*:157–172, on p. 162.

[25] Florian Cajori, *The Teaching and History of Mathematics in the United States* (Washington: Government Printing Office, 1890), p. 71.

[26] Hogan, "Robert Adrain," p. 169.

mous review in which he stated his preference for Newtonian notation:

> Our author evidently used the language of the differential system of Leibnitz, and thus loses the great advantage that attends the genuine fluxions of Newton.[27]

Although Hogan overlooks these seemingly contradictory statements concerning Adrain's acceptance of the new French methods, those very contradictions serve to show the transitional nature of American mathematics in the first few decades of the nineteenth century. Adrain, like many other English-speaking mathematicians, was educated in the older, Newtonian-synthetical system of mathematics. As a gifted mathematician who had read and understood the more modern French-style mathematics, Adrain was able to appreciate some of its advantages. He was not ready, however, to discard completely his mathematical upbringing to embrace the new innovations in mathematics. This reluctance on the part of Adrain makes the work of Nathaniel Bowditch, John Farrar, Charles Davies and other American mathematicians of the first half of the nineteenth century even more remarkable, as they attempted to bring the newest mathematics to the United States.

6.3 A Growing Interest in French Mathematics: The Influence of Jeremiah Day

A careful analysis of the mathematical works printed in America in the first half of the nineteenth century reveals interesting patterns in mathematical education in the United States.[28] Numerically,

---

[27] *Ibid.*, p. 170. Quoted from *Mathematical Diary*, 1825, *1*:20.

[28] The analysis that follows is based on the exhaustive bibliography by Louis Karpinski, *Bibliography of Mathematical Works Printed in America Through 1850* (New York: Arno Press, 1980).

Decade	Works	Increase
1800s	82	
1810s	102	24%
1820s	150	47%
1830s	244	63%
1840s	325	33%

Table 6.1: Mathematical Works in the United States by Decade. Source: Louis Karpinski, *Bibliography of Mathematical Works Printed in America Through 1850*.

we find steady growth in the number of published mathematical works throughout the period, with the largest growth occurring during the decade of the 1830s.[29]

In addition to the growth of total mathematical volumes, the relative sophistication of the mathematical works appears to have increased through the period. For instance, of the 203 mathematical works printed between 1800 and 1825, 54 percent were concerned with arithmetic while only 7 percent addressed algebra, trigonometry and calculus. However, of the 623 works printed between 1825 and 1850, the percentage of arithmetics dropped to 44 percent while the volumes addressing algebra, trigonometry, and calculus rose to 11 percent. In particular, during the first quarter of the century only two books concerned primarily with calculus are listed in Karpinski's bibliography, while in the second quarter of the century the number had risen to nine.[30]

A majority of the early works published in this period were reprints of mathematical textbooks by British authors. In fact,

---

[29] It should be noted that these numbers represent all listings in Karpinski's work, without regard to their mathematical sophistication. Many of these publications were mathematical tables and games, or works on natural philosophy or education that may have had little actual mathematical content.

[30] Included in the nine calculus books is James Ryan's *The Differential and Integral Calculus* (1828), the first calculus textbook written by an American.

the first calculus text printed in the United States was a reprint of Vince's *The principles of fluxions* in 1812. However, the growing list of calculus texts through the first half of the century, many of which followed the continental style of calculus, indicates an American awakening to French mathematics and a slow transition from mathematics dominated by the synthetic style to one emphasizing analysis.

Two years after the publication of Vince's *Fluxions*, Jeremiah Day, professor of mathematics and natural philosophy at Yale from 1801 to 1817, published *An introduction to algebra, being the first part of a Course in mathematics, adapted to the method of instruction in the American colleges*. This work, the first *Algebra* by an American, went through more than sixty-seven editions.[31] Day's *Algebra* is seen by many as an important transitional work. Although written in the British synthetic style,[32] Day's *Algebra* was actually patterned after several British texts that were themselves patterned after works by Euler and Lacroix.[33] Guralnick views Day's *Algebra* as the dawn of the revolution in American mathematics that was completed by the translation of the French corpus of texts.[34]

In addition to his *Algebra*, Day wrote several other popular textbooks, many of which exhibited a marked French influence. Hogan confirms the opinions of Karpinski and Guralnick when he states:

---

[31] Karpinski calls Day's work "the foremost American algebra for fifty years," Karpinski, *Bibliography*, p. 200.

[32] Pycior, "British Synthetic," p. 125.

[33] Stanley M. Guralnick, *Science and the Ante-Bellum American College* (Philadelphia: The American Philosophical Society, 1975), p. 51.

[34] *Ibid.*

Figure 6.3.1: Jeremiah Day

"Day's texts are also closer to the French than to the English texts in style and approach."[35]

Jeremiah Day later became president of Yale where he was co-author of the influential and controversial *Yale Report*. Day believed mathematics was a central aspect of a liberal education, critical to the training of the mind:

> The study of mathematics, by consent of the ablest men who have been conversant with the business of instruction, is especially adapted to sharpen the intellect, to strengthen the faculty of reason, and to induce a general habit of mind favorable to the discovery of truth and the detection of error.[36]

In addition, Day addressed the question of why the study of mathematics was important to students who had no plan to involve themselves in professions requiring mathematics or science. His answer was two-fold. The first was so the student might judge the value of work done by others in the field; and the second was that although the student might forget the details of his mathematical education, "he still knows where to apply for information, and how to direct his inquiries."[37] These ideas concerning the importance of the study of mathematics to a liberal education were to come into conflict with mathematicians of the new generation who were interested in making mathematics a community of experts gathered in a common profession.

Day was certainly not alone in his belief that mathematics was an essential ingredient in the full development of the youthful mind.

---

[35] Edward R. Hogan, "Theodore Strong and Ante-Bellum American Mathematics," *Historia Mathematica*, 1981, *8*:439–455, on pg. 441. For example, Day placed less emphasis on memorization of rules and more on working examples to illustrate the methods. Day also used Legendre's *Geometry* and Lacroix's *Algebra* as references in his texts.

[36] Reprinted in the *American Journal of Science and Arts*, 1829, *15*:326.

[37] *Ibid.*, p. 327.

Indeed, the desire to make proper use of modern mathematics for just this purpose was a central motivation for the translation movement that became an integral part of American mathematics beginning in the second decade of the nineteenth century. Jasper Adams echoed this commonly held educational philosophy in his review of John Farrar's "Cambridge course of Mathematics":

> A course of mathematical instruction, and, indeed a course of instruction of any other kind, may be considered in two points of view; first in relation to the development of the faculties of the mind, and secondly, as furnishing results to be used for the practical purposes of life...it is particularly with a view to the development of the mental powers, that a course of mathematics is important.[38]

Adams continued his assault on those who would minimize the importance of mathematics in a liberal education:

> Granting, if it is possible, that the Physician, the Divine, the Advocate, or the Judge, may have forgotten every proposition in Geometry, and every principle in Algebra; still he may be indebted to these sciences learned in early life, for no small part of his skill in separating error from truth, for his power of fixed attention, for his caution in admitting proof and in drawing conclusions for the general discipline of his mental facilities, and his capacity for arranging all the parts of a long argument, so that it may result in the clear establishment of the desired truth. Such a habit of mind constitutes *true learning*, a rare acquirement; and ought to be the ultimate object of every system of education. It is capable of application to every subject, at all times, and in every situation. Without the

---
[38] *American Journal of Science and Arts*, 1822, 5:308.

accomplishment of this object, no education can be in a considerable degree complete, much less can the mind be highly cultivated.[39]

The two advantages presented by the study of mathematics—training the mind and practical applications—played a central role in the movement to adopt the French style of mathematics in the United States.

Jeremiah Day's influence extended to the next generation of mathematicians through his student, Theodore Strong.[40] Strong, in turn, helped to bridge the gap between two generations represented by Bowditch and Adrain on the one hand and Benjamin Peirce on the other. Like Bowditch, Theodore Strong was influential in the process of the American adoption of French mathematical methods. Hogan has shown that Strong, who initially exhibited little interest in continental mathematics, eventually became "one of the leaders in introducing Continental mathematics to his countrymen."[41] Strong was among the first Americans to use Leibnizian notation consistently in his works on calculus. Gradually Strong's interests evolved from the synthetic methods of geometry to a more modern interest in analytic geometry.[42]

6.4 Pioneer of the Translation Movement: John Farrar

Two men played crucial roles in bringing the French methods to America through their work as translators. As noted in Chapter 4, Nathaniel Bowditch's translation of Laplace's *Mécanique Céleste*

---

[39] *Ibid.*, pp. 308–9.

[40] Strong's importance was discussed at length in Chapter 3 and Chapter 5.

[41] Hogan, "Theodore Strong," p. 445.

[42] *Ibid.*, p. 447.

marked a high point in American scholarship, enabling generations of astronomers and mathematicians to study the work in English and with the help of his commentary.

The second pioneer in the translation movement was John Farrar, longtime Hollis Professor of Mathematics and Natural Philosophy at Harvard and translator of a series of mathematics and natural philosophy textbooks. Farrar and Bowditch were sometimes called on to work together at Harvard, such as when both were appointed to serve on a committee in 1815 to investigate the possibility of building an observatory at Harvard. Unfortunately, the project died before it could really get started due to a general lack of funding.

Farrar, whose first three translations included algebra texts by Leonhard Euler and Sylvester Lacroix, and an arithmetic text by Lacroix, pioneered the translation movement in American mathematics. Farrar opened the floodgates of American translations of French mathematical works and cultivated an atmosphere in American mathematics that would culminate in the work and teachings of a new generation of mathematicians.

John Farrar, born in Lincoln, Massachusetts in 1779, obtained his B.A. from Harvard in 1803.[43] He became a tutor in Greek at Harvard in 1805 and received an M.A. in 1806. Farrar was appointed Hollis Professor of Mathematics and Natural Philosophy at Harvard in 1807, but only after Nathaniel Bowditch and the Reverend Joseph Mckeen each were offered, and refused, the position. Farrar remained at Harvard as Hollis Professor until 1836. Although not a gifted mathematician, Farrar influenced a genera-

---

[43] The most complete biography of Farrar is his obituary, written by John G. Palfrey and printed in *The Christian Examiner and Religious Miscellany*, 1853, *55*:121–136. See also Ackerberg, *Mathematics*.

Figure 6.4.1: John Farrar

tion of mathematics students at Harvard as an inspiring teacher,[44] according to a testimonial from a former student:

> He delivered, when I was in college, a lecture every week to the Junior class on natural philosophy, and one to the Senior class on astronomy. His were the only exercises at which there was no need of a roll-call. No student was willingly absent. The professor had no notes, and commenced his lecture in a conversational tone and manner, very much as if he were explaining his subject to a single learner. But whatever the subject, he very soon rose from prosaic details to general laws and the principles, which he seemed ever to approach with blended enthusiasm and reverence, as if he were investigating and expounding divine mysteries. His face glowed with inspiration of his theme. His voice, which was unmanageable as he grew warm, broke into a shrill falsetto; and with the first high treble notes the class began to listen in breathless stillness, so that a pin-fall could, I doubt not, have been heard through the room. This high key once reached there was no return to the lower notes, nor any intermission in the outflow and quickening rush of lofty thought and profound feeling, till the bell announced the close of the hour, and he piled up all the meaning that he could stow into a parting sentence, which was at once the climax of the lecture, and the climax of an ascending scale of vocal utterance higher, I think, than is within the range of an ordinary soprano singer. I still remember

---

[44] Cohen calls Farrar "one of the most inspired teachers and lecturers ever to grace a Harvard lecture platform." I.B. Cohen, *Some Early Tools of American Science* (New York: Russell and Russell, 1967), p. 21.

portions of his lectures, and they now seem to me no less impressive than they did in my boyhood.[45]

Farrar was trained as a minister[46] (as were the majority of college graduates at the time) and believed the natural world was best understood as a work of God. His former student continued:

> I recall distinctly a lecture in which he exhibited, in its various aspects, the idea that in mathematical science, and in it alone, man sees things precisely as God sees them, handles the very scale and compasses with which the Creator planned and built the universe; another in which he represented the law of gravitation as coincident with, and demonstrative of, the divine omnipresence; another, in which he made us almost hear the music of the spheres, as he described the grand procession, in infinite space and in immeasurable orbits, of our solar system and the (so called) fixed stars, His lectures were poems, and hardly poems in prose; for his language was unconsciously rhythmical, and his utterances were like a temple chant.[47]

Farrar's influence as a teacher is further illustrated by the fact that in his twenty-nine years at Harvard, there were 275 theses on mathematical subjects written by Harvard students.[48] Many of these theses were on the subject of fluxions and/or differential

---

[45] Andrew P. Peabody, *Harvard Reminiscences* (Boston, 1888), p. 70. Quoted in Cajori, *Teaching and History*, pp. 127–128.

[46] In fact, Farrar was well on his way to a career in the pulpit when he was unexpectedly offered the position of Tutor in Greek at Harvard in 1805. It was only the persuasion of his brother that convinced Farrar to accept the position and begin his career at Harvard. Palfrey, "John Farrar," p.125.

[47] Cajori, *Teaching and History*, p. 128.

[48] *Ibid.*, p. 132.

calculus. Among the students who wrote such theses were future luminaries like George Bancroft, George Emerson, Warren Colburn, Sears Cook Walker, Benjamin Peirce, and Wendell Phillips.[49]

The influence of Farrar is also evident through an examination of the mathematical curriculum at Harvard. In 1802, freshmen at Harvard studied arithmetic, sophomores continued with more arithmetic and added some algebra, juniors studied Euclid, trigonometry, and conic sections and seniors studied spheric geometry and trigonometry.[50] This mathematics curriculum was essentially unchanged since 1788, when the regulations governing the Hollis Chair of Mathematics and Natural Philosophy were established. In addition, the Hollis Chair was to provide instruction to any student who might want to "pursue the study of Fluxions or any other abstruse parts of the Mathematics."[51] By 1830, arithmetic was an entrance requirement at Harvard.[52] This allowed freshmen

---

[49] *Ibid.*

[50] Russel Blaine Nye, *The Cultural Life of the New Nation* (New York: Harper and Row, 1960), p.188.

[51] Cohen, *Early Tools*, pp. 57–58.

[52] One indication of the changes made to entrance requirements in one generation comes from a description of Farrar's own admission examination, which makes no mention of mathematics:

> We took our books and set off, walked across the Common; and ascended four or five long winding stairs into a huge, dark, hot entry, and there waited until half past eleven, which was about an hour and a half, when the division that had been under examination removed to the Museum. We were called in, took our seats before Messrs. Barron, Hedge, Wells, and Kendall, and our names, ages, &c. were taken off. Then Mr. Kendall examined us in the sixth chapter to the Romans, Mr. Wells in the tenth section "Pro T. Annio Milone," and then we were dismissed until half past two, at which time we were returned and waited until after five, when we were again called for. We took our places, and were examined in the fifth book of Virgil. We were then sent into the Museum to make Latin, which being examined, we were called in. Our names all being called, we were sent to the President, "not a little comforted." He was so unwell, he only looked over our Latin, and then said: "I accept you all," giving us an extract from the laws. We went to the steward, got our bond, and so got through about eight of the clock.

Palfrey, "John Farrar," p.123.

to begin with plane geometry, algebra, and solid geometry; sophomores were taught trigonometry, topography, and calculus; juniors studied natural philosophy, mechanics, electricity, and magnetism; and seniors progressed to optics and natural philosophy.[53] During Farrar's tenure at Harvard,

> The chief labor and the crowning honor of successful scholarship were in mathematics and the classics. The mathematical course extended through the entire four years, embracing the differential calculus, the mathematical treatment of all departments of physical science then studied, and a thoroughly mathematical treatise on astronomy.[54]

In spite of his reputation as an inspiring teacher, Farrar's most important contribution to American mathematics was the foundational role he played in the gradual replacement of an old system of mathematics based on British synthetic style with the new system of French analytical mathematics. Farrar made a major contribution to this transition with his series of translations of French textbooks. In addition, Farrar translated several natural philosophy texts that changed the way science was taught in America. His mathematical texts became known as the "Cambridge Course of Mathematics" while his other works on electricity and magnetism, mechanics, optics, and astronomy were known as the "Cambridge Course of Natural Philosophy."

Farrar's first translation was Leonhard Euler's *An introduction to the elements of algebra, designed for the use of those who are acquainted only with the first principles of arithmetic*, which first appeared in 1818. It went through a total of four editions, the

---

[53] Cajori, *Teaching and History*, p. 132.

[54] Quoted in *Ibid.*, p. 133.

last appearing in 1836.[55] Farrar's translation of Euler's *Algebra* was the first foreign language mathematics book translated and used as a text in America.[56] The second book in the "Cambridge Course of Mathematics," also appearing in 1818, was a translation of Sylvestre Lacroix's *An elementary treatise on arithmetic.* This text also went through a total of four editions, the last appearing in 1834.

Farrar made adaptations to Lacroix's *Arithmetic*, as he did with all of his translations. He adapted the text to the American system of weights and measures, as the French had adopted the metric system. He also adapted Lacroix's text to the American system of currency. Finally, Farrar deleted some sections of the text and added others in an attempt to make the *Arithmetic* more accessible to Harvard students. When these texts were first published, they were taught to students at Harvard. By 1822, Lacroix's *Arithmetic* and Euler's *Algebra* were required for admission to Harvard.[57]

Finishing a busy year, Farrar published his translation of Lacroix's *Elements of algebra* in 1818. This text went through five editions, the last appearing in 1837. Lacroix's *Algebra* was more advanced than had been Euler's, covering subjects such as the theory of equations, infinite quantities, the binomial theorem, and logarithms. A review of Farrar's translation stated, "The translation is performed with ability and fidelity," containing few translation errors and including numerous valuable explanatory notes added by the translator.[58] Farrar intended Lacroix's *Algebra* for students at

---

[55] See Appendix 4 for a complete list of Farrar's translations of mathematical texts.

[56] Lao G. Simons, "The Influence of French Mathematicians at the end of the Eighteenth Century upon the Teaching of Mathematics in American Colleges," *ISIS*, 1931, *15*:104–123, on pg. 114.

[57] Adams, "Review," p. 307.

[58] *Ibid.*, p. 325.

Harvard who had already mastered the more elementary material in Euler's *Algebra*.

Elements of Geometry—a translation of Adrien-Marie Legendre's work and published by Farrar in 1819—was one of Farrar's most successful texts, appearing in nine editions, the last in 1841. Farrar's reviewer once again praised the accuracy and value of the translation, but lamented the fact that Farrar had not chosen to translate the entire work:

> We are persuaded, that all lovers of mathematical learning, after having perused the translation, will feel induced to go to the expense of sending out for the original, for the sake of those parts, which the translation does not contain.[59]

In this work we find some hints as to Farrar's motivations for undertaking his task. In the "Advertisement" by Farrar at the beginning of the text, he noted that Legendre's *Geometry* was chosen because it was "thought to unite the advantages of modern discoveries and improvements with the strictness of the ancient method."[60] This statement highlights Farrar's position in a transitional generation. He saw importance in teaching the students at Harvard the newest discoveries in mathematics, yet retained a desire for the time-tested synthetic proofs of the Greeks.[61] Farrar had one foot in the new analytic camp and another in the traditional synthetic one.

---

[59] *Ibid.*, p. 301.

[60] John Farrar, *Elements of Geometry* (Cambridge, MA: Hilliard and Metcalf, 1819), Advertisement.

[61] Farrar's desire to honor the ancient synthetic methods is actually a reflection of Legendre's own words in prefacing his work: "The method of the ancients is very generally regarded as the most satisfactory and the most proper for representing geometric truths...it offers...a discipline of peculiar kind, distinct from that of analysis." *Ibid.*, p. v.

Two other characteristics common to this generation of American mathematicians can be seen in Farrar's next comments:

> It [Legendre's *Geometry*] has now been in use for a considerable number of years, and its character is sufficiently established. It is generally considered as the most complete and extensive treatise on the elements of geometry, which has yet appeared. It has been adopted as the basis of the article on geometry in the fourth edition of the Encyclopedia Brittanica, lately published, and in the Edinburgh Encyclopedia, edited by Dr. Brewster.[62]

First of all, we see that Farrar was not breaking new ground in adopting this French work. He even seemed somewhat hesitant to do so. He justified his adoption of Legendre's *Geometry* by pointing out that the work was already well established. Secondly, although Farrar's translations signify a shift away from the traditional British style of mathematics, Farrar deemed it important to note that even this work had been well received and used extensively by British mathematicians. The "break" from British influence may not be as sharp as it first appears.

Farrar also addressed the preparedness of the students he intended to reach with the translation. He omitted several sections of Legendre's original work, feeling these sections were "less useful" than the others or they "require[d] more attention than other parts of the work."[63] Furthermore, it is evident that the students at Harvard were generally less prepared than Legendre's audience; therefore, Farrar felt the need to add a section to prepare his students for their studies:

---

[62] *Ibid.*, Advertisement.

[63] *Ibid.*

> As the reader is supposed to be acquainted with algebraical signs and the theory of proportions, brief explanation of these, taken chiefly from Lacroix's geometry, and forming properly a supplement to his arithmetic, is prefixed to the work under the title of an introduction.[64]

In this short introduction, Farrar defined some common algebraic symbols, provided a short discussion of powers, and presented a concise discussion of the theory of proportions. All of these subjects would have been part of even a rudimentary training in mathematics. That Farrar felt the need to include them in his text indicated that at least some of the incoming students at Harvard did not have this training.

The modest success Farrar found with his translation was not, however, comparable to that of another translation of Legendre's *Geometry* made five years later. The British physicist David Brewster commissioned Thomas Carlyle, a former student at Edinburgh University who would later become a renowned writer, to translate Legendre's work. Published in 1824, Brewster's edition of Legendre, modified by Charles Davies for American audiences, continued to be issued in the United States until 1885. This translation also played a central role in a long-standing debate in Britain concerning geometry versus the new methods of analysis.[65] Craik maintains that Carlyle was unaware of Farrar's translation at the time of his own work on Legendre's *Geometry*, but it is unclear whether or not Brewster was aware of Farrar's work.[66]

Carlyle's ignorance of Farrar's work raises the question of Farrar's reputation abroad. Although certainly not as well known or

---

[64] *Ibid.*

[65] See Alex D.D. Craik, "Geometry versus Analysis in Early 19th-Century Scotland: John Leslie, William Wallace, and Thomas Carlyle," *Historia Mathematica,* 2000, *27*:133–163.

[66] *Ibid*, p. 147.

respected in Europe as Nathaniel Bowditch, it does appear that European scientists knew of Farrar. In at least one visit abroad, Farrar was received by European scientists. Farrar's widow related one such visit, prompted by a letter from "the Professor of Mathematics in Cambridge" who wrote to Farrar, "Come to Cambridge; you need bring no letters of introduction; we all know you, and want to see you."[67] During this visit, Farrar and his wife were also guests of Mary Somerville, where Farrar and Somerville "talk[ed] away on the higher branches of mathematics and astronomy."[68]

Farrar continued his mathematical series in 1820 with *An elementary treatise on plane and spherical trigonometry, and on the application of algebra to geometry; from the mathematics of Lacroix and Bézout*. This text went through four editions, the last appearing in 1837. In this text, the section on trigonometry was taken from Lacroix's *Course of Mathematics* while the section on applications of algebra to geometry was from Étienne Bézout's *Algebra*. Once again, Farrar chose a traditional text in Bézout. He mentioned that he considered a more advanced and newer text such as Lacroix or Biot,

> but as analytical geometry has hitherto made no part of the mathematics taught in the public seminaries of the United States, and as only a small portion of time is allotted to such studies, and this in many instances at an age not sufficiently mature for inquiries of an abstract nature, it was thought best to make the experiment with a treatise distinguished for its simplicity and plainness.[69]

---

[67] Mrs. John Farrar, *Recollections of seventy years* (Boston: Tiknor and Fields, 1865), p. 184.

[68] *Ibid.*, p. 186.

[69] John Farrar, *Plane and spherical trigonometry, and on the application of algebra to geometry; from the mathematics of Lacroix and Bézout*, 3rd edition (Boston: Hilliard, Gray and Co., 1833), Advertisement.

As he had in his translation of Legendre's *Geometry*, Farrar supplied extra explanation and references for the students who were not adequately prepared to read the text. Once again we see that Farrar prepared his translation with the special needs of American students in mind.

In 1822, Farrar published *An elementary treatise on the application of trigonometry to orthographic and stereographic projection, dialing, mensuration, navigation, nautical astronomy, surveying and leveling; together with logarithmic and other tables; designed for the students of the University at Cambridge*, which went through four editions, the last in 1840. This text assumed the student had studied trigonometry, probably from Farrar's translation of Lacroix. Farrar used Bonnycastle's *Trigonometry*, Jean-Baptiste Delambre's *Astronomy*, Bézout's *Navigation*, and Louis Puissant and Malortie's *Topography* to compile the text, quite an eclectic mix. Most of the tables came from Bowditch's *Practical Navigator*. Although the title sounds as if the text emphasizes application, Farrar contended that the texts used in America on this subject were too practical in nature, and his work would be more useful for someone in "pursuit of liberal studies."[70]

In *Applications of Trigonometry*, Farrar announced that the last book in the series "Course of Mathematics," his treatise on calculus, was nearly ready for the press. He also announced his intention to begin efforts to publish a work on natural philosophy, which turned out to be an entire series known as the "Course of Natural Philosophy."[71]

The last text in the "Cambridge Course of Mathematics," *First principles of the differential and integral calculus, or the doctrine of fluxions*, appeared in 1824. This text, translated from Bézout, went

---

[70] *Ibid.*, Advertisement.

[71] *Ibid.*

through only 2 editions, the last in 1836. Bézout's *Cours de Mathématiques à l'usage des Gardes du Pavillon et de la Marine* was a highly influential multi-volume work published in the later part of the eighteenth century in Paris. Farrar translated the first part of volume four of this series, *Les Principes généraux de la Méchanique, précédés des Principes de Calcul qui servent d'introduction aux Sciences Physico-Mathématiques,* originally published by Bézout in 1795.

Farrar's inclusion of "The doctrine of fluxions" in the title of his text is interesting. Bézout's original work was one that strictly adhered to the differential and integral notation of Leibniz and of contemporary French mathematics. Surprisingly (at least viewed in light of Farrar's chosen title), Farrar's English translation retained the Leibnizian notation with no attempt to convert anything to fluxional notation. Farrar did end his text with some notes on such things as a demonstration of the binomial formula, the method of exhaustion, the method of indivisibles, and the (Newtonian) method of prime and ultimate ratios. Although these subjects were historically associated with older methods in calculus, and in the case of prime and ultimate ratios with fluxions, these short notes would not seem to justify including fluxions in the title.

There are several possible explanations for Farrar's inclusion of the term fluxions in the title. This may be further evidence that he and other American mathematicians were not ready to sever their ties to the British traditions in mathematics completely. Farrar and others realized the advantages of the new methods of analysis developed by the French, but retained ideas that might be called provincial concerning the discovery of calculus and the foundations upon which it was built.

It is also possible that Farrar, as a man who was attempting to change the American landscape in calculus, realized that he might need a "hook," such as the word fluxions in his title, to catch the attention of the reader and make the transition using terms that were familiar to American mathematics students.[72] Or the explanation might be of an even simpler nature. Perhaps Farrar's use of the word fluxion only signified the equivalence of the two methods to an audience more familiar with the Newtonian terminology. Whatever the case, we have seen in Chapter 3 that the intermingling of Newtonian and Leibnizian notation was becoming a common phenomenon in American mathematical publications.

Although to some extent outside the scope of this study, mention should be made of Farrar's efforts in forming a "Cambridge Natural Philosophy Series." His series of translations and original texts were the primary texts used (along with several by Denison Olmsted of Yale) at American colleges for some twenty years.[73] It should be noted that the title page of each translation includes "for the use of the students of the university at Cambridge, New England." Just as he had in his mathematics series, Farrar strove to adapt the natural philosophy series to the abilities and interests of American students.

Historians differ in their opinions as to the importance of John Farrar to American mathematics. Brooke Hindle credits Farrar as being primarily responsible for the shift from Newtonian to Leibnizian notation in calculus in America.[74] Helena Pycior sees Farrar's translations as part of a bigger picture at Harvard of a growing appreciation of French mathematical superiority, an approach

---

[72] Newton's fluxions were not easily forgotten. As late as 1879, American textbook authors were using fluxions as a handy tool to teach related rates problems in calculus: "Our plan is to return to the method of fluxions, and making use of the precise and easily comprehended definitions of Newton." John Minot Rice and William Woolsey Johnson, *An Elementary Treatise on the Differential Calculus Founded on the Method of Rates or Fluxions*. Quoted in Bill Austin, Don Barry and David Berman, "The Lengthening Shadow: The Story of Related Rates," *Mathematics Magazine*, 2000, 73:3–12, on pg. 8.

[73] Guralnick, *Science*, p. 66.

[74] Brooke Hindle, "John Farrar," in Charles Coulston Gillispie, ed., *The Dictionary of Scientific Biography* (New York: Scribner, 1981), vol. 4, pp. 546–547.

she largely credits to Nathaniel Bowditch.[75] Cajori agrees, calling Nathaniel Bowditch's translation of Laplace's *Mécanique Céleste* an important stimulus to the study of French mathematics in America.[76]

Bowditch's influence is problematic, however, when one remembers that his translation of Laplace's *Mécanique Céleste* did not appear until the first volume was published in 1829. By this time, Farrar's textbooks had been employed for many years at Harvard, and Charles Davies was beginning to wield an even greater influence with his series of mathematical texts. Benjamin Peirce, generally considered America's first research mathematician, ascribed to Farrar the honor of being the most important person in "the adoption of the present admirable system of instruction in the mathematical sciences [in America]."[77] This opinion was given in spite of the fact that Peirce was often critical of Farrar's texts.

Peirce's criticism of Farrar's translations, and the criticism Farrar received from others, was directed at his habit of choosing already out-of-date French texts for his translations. His choice of texts, however, was based upon a belief that lack of preparation and interest on the part of American students made it essential that more elementary and easier to understand texts be translated.[78] Rosenstein points out as American mathematicians became better educated in the newest mathematics, they grew less likely to accept earlier texts and instead looked for those with the solid logical foundations supplied by Cauchy and Weierstrass.[79]

---

[75] Pycior, "British Synthetic," pp. 137–138.

[76] Cajori, *Teaching and History*, p. 104–105.

[77] Quoted in Simons, "Influence," p. 111.

[78] Cajori, *Teaching and History*, p. 130.

[79] George M. Rosenstein, "The Best Method. American Calculus Textbooks of the Nineteenth Century," in Peter Duren, ed., *A Century of Mathematics in America*, Volume 3 (Providence, RI: American Mathematical Society, 1988), pp. 77–109, on p. 101.

Charles Davies was a leader in this second phase of the translation movement, choosing newer French authors as his objective. Davies was the first American to use limits in his lectures and wrote the first commercially successful calculus text in the United States, although his text was still based upon a French text by Boucharlat.[80] Davies completed the work that Farrar began. By 1836, all American colleges offered courses derived from the French mathematical style.[81] It is unfair, however, to criticize Farrar for not choosing the same texts as did Davies, for Farrar was of an earlier generation. Farrar's importance lies in the fact that he was a pioneer in the American mathematical efforts to adopt the French style. As Pycior notes,

> Day and Farrar belonged to the first generation of Americans exposed to the rival algebraic styles; Davies and Peirce, to the second generation.[82]

In spite of the criticisms of Farrar's translations, his work laid the foundation for future generations of American mathematicians. His series of books addressing mathematics and natural philosophy was a first step in the process of exposing American students to the ideas of continental European scientists. Farrar's series of translations were used widely at American universities, if only for a short period of time. Even at West Point in the 1820s, where Charles Davies was soon to begin his own successful textbook series, Farrar's texts were an integral part of the mathematics curriculum. In 1825, for instance, Farrar's translations of Lacroix's *Elements*

---

[80] *Ibid.*, p. 83.

[81] Guralnick, *Science*, p. 54.

[82] Pycior, "British synthetic," p. 145.

of Algebra, Legendre's *Geometry* and the *Trigonometry* taken from Lacroix and Bézout were used at the Academy.[83]

6.5 Charles Davies and West Point: Emulating the École Polytechnique

Almost from its inception, the Military Academy at West Point played a leading role in integrating French learning in mathematics, natural philosophy and engineering into the American college curriculum. By the time Sylvanus Thayer became superintendent at West Point in 1817, the Academy had already established a history of employing the best available instructors in mathematics and the sciences. George Baron, the English immigrant who established the first mathematical journal in the United States, was also the first mathematics instructor at the Military Academy from 1801 to 1802. He was followed by Jared Mansfield, who published an early American work in natural philosophy called *Essays, Mathematical and Physical* in 1802. Later mathematics instructors at the Academy included Ferdinand Hassler, the first head of the United States Coast Survey,[84] Andrew Ellicott, an important figure in American science for many decades, and Charles Davies, whose leadership in teaching and translating the newest French mathematical works would influence several generations of American mathematicians.

The year 1816 was especially eventful at West Point, and it proved to be important for American science at large. In 1816

---

[83] Joe Albree, David C. Arney and V. Fredrick Rickey, *A Station Favorable to the Pursuit of Science* (Providence, RI: American Mathematical Society, 2000) pp. 14–15.

[84] Hassler actually attempted to introduce French mathematics at West Point from 1807–1810, but was unsuccessful primarily due to the general chaotic atmosphere at the Academy and the poor preparation of the students. John C. Greene, *American Science in the Age of Jefferson* (Ames: The Iowa State University Press, 1984), p. 131.

William McRae and Sylvanus Thayer traveled to France to study the methods of instruction at the École Polytechnique in hopes of patterning the Military Academy at West Point after the French military academy. McRae and Thayer returned with over 1000 books for the West Point library.[85] Among these books were the newest in Continental mathematics, including works by Monge, Legendre, Euler, Lacroix, Montucla, and Wronski.[86] As superintendent of the Academy, Thayer continued to build the Military Academy's mathematical collections through the 1820s, adding classics in mathematics to the library and acquiring full runs of the *Journal de l'Ecole Polytechnique*, the *Annales de Mathématiques Pures et Appliquées* and the *Transactions of the Royal Society*.[87] To demonstrate further the influence of the French on West Point, by 1830 French titles composed 34 percent of the mathematics collection at the Academy's library.[88]

It was also in 1816 that Charles Davies was appointed assistant professor of mathematics at the Academy, becoming a full professor in 1823 and staying at West Point until 1837. Davies, along with engineering professor Claudius Crozet, began teaching the methods of the French and other continental mathematicians. Crozet introduced descriptive geometry to the cadets at West Point in 1817. He attempted to teach the new mathematical methods to his engineering classes, in spite of the fact the mathematics instructors like Ellicott and Mansfield were still teaching British-style mathemat-

---

[85] Albree, *Station*, p. 1.

[86] *Ibid.*, p. 24.

[87] *Ibid.*, p. 25.

[88] *Ibid.*, p. 26. This represents a substantial change in direction for the Military Academy's mathematics collection. The catalog of books belonging to the Academy in 1803 (Albree, pp. 235–237) shows an almost entirely British influence in the small collection of mathematics books.

ics.[89] The program of technical education at West Point influenced future scientific schools founded in 1846 at Yale (later named the Sheffield Scientific School) and in 1847 at Harvard (the Lawrence Scientific School). In the process, West Point's program supplied the first engineering professors at these and other American colleges.[90]

6.6 Contributing Factors to the Adoption of a New System of Mathematics

Why did American mathematicians awaken to French mathematics in the first third of the nineteenth century? Several reasons may be found. First, after the War of 1812, Americans possessed a newfound patriotism and an urgent need to make a clean break from Britain. Americans were traveling to Europe for educational purposes[91] and European scholarship, first French and later German, was beginning to exhibit marked influence on American culture. Secondly, a "revolution" was occurring in higher education in the United States, with a call for more emphasis on liberal arts and science and less emphasis on classical languages.[92] The Yale Report of 1828 was symptomatic of this trend in American higher education. Because of the changes occurring in higher education, the

---

[89] Greene, *American Science*, p. 131.

[90] Albree, *Station*, p. 22.

[91] Travel to Europe was easier and more common, even more so with the advent of trans-Atlantic steamship travel in 1838. The trend towards European travel among American scientists became pronounced in the later 1830s. See Bruce Sinclair, "Americans Abroad: Science and Cultural Nationalism in the Early Nineteenth Century," in Nathan Reingold, ed., *The Sciences in the American Context: New Perspectives* (Washington D.C.: Smithsonian Institution Press, 1979), pp. 35–53.

[92] Guralnick, *Science*, p. 33.

American textbook industry stood ready to serve a more sophisticated curriculum. Students were better prepared. Change was accepted as a fact of collegiate life.[93]

Finally, the leaders of American mathematics such as Nathaniel Bowditch and John Farrar spearheaded the drive to integrate continental analysis into the American mathematical curriculum.

The American mathematical community counts as one of its foundations the pioneering efforts of mathematicians such as John Farrar who first brought modern Continental mathematics to the United States. Although his translations of classic French mathematical works were soon outdated and superseded, Farrar made the first serious attempt to educate American students in the methods of analysis. In doing so, Farrar provided the groundwork for much of what was to come in American mathematics.

---

[93] *Ibid.*

Bibliography

Ackerberg, Amy. *Mathematics is a Gentleman's Art: Analysis and Synthesis in American College Geometry Teaching, 1790-1840*. Ph.D. dissertation. Iowa State University, 2000.

Ackerberg-Hastings, Amy. "John Playfair on British decline in mathematics." *BSHM Bulletin: Journal for the British Society for the History of Mathematics* 23 (2008): 81-95.

Adams Family Papers: An Electronic Archive. Massachusetts Historical Society. http://www.masshist.org/digitaladams/

Adams, Jasper. "Review of Cambridge Course of Mathematics" *American Journal of Science and Arts* 5 (1822):304-325, continued in 6 (1823):283-321.

Albree, Joe; Arney, David C.; and Rickey, V. Frederick. *A Station Favorable to the Pursuit of Science*. Providence, RI: American Mathematical Society, 2000.

American Review and Literary Journal 1 (1801).

Anderson, Henry James. "On the motion of Solids on Surfaces, in the Two Hypotheses of Perfect Sliding and Perfect Rolling, with a Particular Examination of their small Oscillatory Motions." *Transactions of the American Philosophical Society* 3 (1830):315-382.

Anonymous. "A Review of the *Principia* of Newton." *American Journal of Science and Arts* 11 (1826):238-245, continued in 13 (1828):311-321.

———. "A Review of 'A Treatise of Mechanics'." *American Journal of Science and Arts* 7 (1824):72-85.

Archibald, Raymond Clare. "The Scientific Achievements of Nathaniel Bowditch." in *A Catalogue of a Special Exhibition of Manuscripts, Books, Portraits, and Personal Relics of Nathaniel Bowditch (1773-1838)*. Salem, MA: Peabody Museum, 1937, pp. 7-16.

———. "Benjamin Peirce." *American Mathematical Monthly* 32 (1925):1-30.

Austin, Bill; Barry, Don; and Berman, David. "The Lengthening Shadow: The Story of Related Rates." *Mathematics Magazine* 73 (2000):3-12.

Bates, Ralph. *Scientific Societies in the United States*. Cambridge, MA: The MIT Press, 1965.

Beckmann, Petr. *A History of Pi*. New York: Barnes & Noble Books, 1971.

Bedini, Silvio A. *Thinkers and Tinkers: the early American men of science*. New York: Scribner, 1975.

Berry, Robert Elton. *Yankee Stargazer: The Life of Nathaniel Bowditch*. New York: McGraw Hill, 1941.

Birkhoff, Garrett. "Mathematics at Harvard, 1836-1944." in Peter Duren, ed., *A Century of Mathematics in America, Part II*. Providence, RI: American Mathematical Society, 1988, pp. 3-58.

Bowditch Collection, Boston Public Library.

Bowditch, Nathaniel Ingersoll. "Memoir of Nathaniel Bowditch." in volume I of LaPlace, Pierre Simon. *Mécanique Céleste*. Translated by Nathaniel Bowditch.

Bowditch, Nathaniel. *The New American Practical Navigator: being an Epitome of Navigation; containing all the Tables necessary to be used with the Nautical Almanac, in determining the Latitude; and the Longitude by lunar observations; and keeping a complete reckoning at sea: illustrated by proper rules and examples:*

the whole exemplified in a Journal kept from Boston to Medeira, in which all the rules of navigation are introduced: also the demonstration of the most useful Rules of Trigonometry: with many useful Problems in Mensuration, Surveying, and Gauging, and a Dictionary of Sea-Terms; with the Manner of performing the most common Evolutions at sea. To which are added, some general instructions and information to Merchants, Masters of Vessels, and others concerned in Navigation, relative to Maritime Laws and Mercantile customs. From the best authorities: Enriched with a number of New Tables, with original improvements and additions, and a large variety of new and important matter: also, many thousand errors are corrected, which appeared in the best systems of navigation yet published. Newburyport, 1799.

———. *Celestial Mechanics*, 4 volumes. Boston, 1832-1834.

———. "New Method of Working Lunar Observations." *Memoirs of the American Academy of Arts and Sciences* 2 (1804):1-11.

———. "On the Motion of a Pendulum Suspended From Two Points." *Memoirs of the American Academy of Arts and* Sciences 3 (1815):413-436.

Bowditch, Susan W. *Nathaniel Bowditch, 1773-1838.* (unpublished manuscript).

Boyer, Carl. *The History of Calculus and its Conceptual Development.* New York: Dover, 1959.

Brooks, Van Wyck. *The Flowering of New England (1815-1865).* New York: Random House, 1936.

Bruce, Robert. *The Launching of Modern American Science, 1846-1876.* New York: Alfred A. Knopf, 1987.

Bullock, Steven C. *Revolutionary Brotherhood: Freemasonry and the Transformation of the American Social Order, 1730-1840.* Chapel Hill: University of North Carolina Press, 1996.

Burnham, John C., ed. *Science in America.* New York: Holt, Rinehart and Winston, 1971.

Cajori, Florian. *The Teaching and History of Mathematics in the United States.* Washington: Government Printing Office, 1890.

——. *The Chequered Career of Ferdinand Rudolph Hassler, first superintendent of the United States Coast Survey.* Boston: The Christopher publishing house, 1929.

Cannell, D.M. "George Greene: An Enigmatic Mathematician." *American Mathematical Monthly* 106 (1999):136-151.

Cannon, Susan Faye. *Science in Culture: the early Victorian period.* New York: Science History Publications, 1978.

Chaplin, Joyce E. *The First Scientific American: Benjamin Franklin and the Pursuit of Genius.* New York: Basic Books, 2006.

Clay, Joseph. "Observations on the Figure of the Earth." *Transactions of the American Philosophical Society* 5 (1802):312-319.

Cleaveland, Parker. "Observations of the eclipse of the sun of Sep. 17, 1811, made at Bowdoin College, Brunswick." *Memoirs of the American Academy of Arts and Sciences* 3 (1815):247-248.

Cohen, I Bernard. *Franklin and Newton: An Inquiry into the Speculative Newtonian Experimental Science and Franklin's Work in Electricity as an Example Thereof.* Philadelphia: The American Philosophical Society, 1956.

——. *Some Early Tools of American Science.* New York: Russell and Russell, 1967.

——. *Benjamin Peirce: Father of Pure Mathematics in America.* New York: Arno Press, 1980.

——. *Science and the Founding Fathers: Science in the Political Thought of Thomas Jefferson, Benjamin Franklin, John Adams, and James Madison.* New York: W.W. Norton and Company, 1995.

Cooke, Roger and Rickey, V. Frederick. "W.E. Story of Hopkins and Clark." in Duren, ed. *A Century of Mathematics*, vol. 3, pp. 71-74.

Coolidge, J.L. "Three Hundred Years of Mathematics at Harvard." *American Mathematical Monthly* 50 (1943):347-356.

——. "Robert Adrain, and the Beginnings of American Mathematics." *American Mathematical Monthly* 33 (1926):61-76.

Craik, Alex. "Geometry versus Analysis in Early 19th-Century Scotland: John Leslie, William Wallace, and John Carlyle." *Historia Mathematica* 27 (2000): 133-163.

Daniels, George H. *American Science in the Age of Jackson.* New York: Columbia University Press, 1968.

Dean, James. "Of the apparent motion of the earth viewed from the moon, arising from the moon's librations." *Memoirs of the American Academy of Arts and Sciences* 3 (1815):241-245.

Despeaux, Sloan Evans. *The Development of a Publication Community: Nineteenth-Century Mathematics in British Scientific Journals.* Ph.D. dissertation, University of Virginia, 2002.

Dictionary of American Biography. New York: Charles Scribner's Sons, 1996.

Dupree, A. Hunter. *Science in the Federal Government: A History of Policies and Activities to 1940.* Cambridge, MA: Belknap Press, 1957.

Duren, Peter, ed. *A Century of Mathematics in America*, 3 vols. Providence, RI: The American Mathematical Society, 1988.

Edwards, C.H. *The Historical Development of the Calculus.* New York: Springer-Verlag, 1979.

Eisele, Carolyn. "Benjamin Peirce." *Dictionary of Scientific Biography,* vol. 10, pp. 478-480.

Ellicott, Andrew. "Astronomical Observations, &c." *Transactions of the American Philosophical Society* 1(new series) (1818):93-102.

Elliott Clark.*Biographical Dictionary of American Science: the Seventeenth Through the Nineteenth Centuries.* New York: Greenwood Press, 1979.

Enros, P.J. "The Analytical Society (1812-1813): Precursor of the renewal of Cambridge mathematics." *Historia Mathematica* 10 (1983):24-47

Farrar, John, trans. *An introduction to the elements of algebra, designed for the use of those who are acquainted only with the first principles of arithmetic.* Leonhard Euler. Cambridge, 1818.

———. *An elementary treatise on arithmetic.* Silvestre Lacroix. Boston, 1818.

———. *Elements of Algebra.* Silvestre Lacroix. Cambridge, 1818.

———. *Elements of Geometry.* Adrien-Marie Legendre. Cambridge, 1819.

———. *An elementary treatise on plane and spherical trigonometry, and on the application of algebra to geometry; from the mathematics of Lacroix and Bézout.* Silvestre Lacroix and Etienne Bézout. Cambridge, 1820.

———. *An elementary treatise on the application of Trigonometry to Orthographic and Stereographic Projection, Dialling, Mensuration of Heights and Distances, Navigation, Nautical Astronomy, Surveying and Levelling; together with Logarithmic and Other Tables; designed for the Use of the Students of the University at Cambridge, New England.* Cambridge, 1822.

———. *First principles of the differential and integral calculus, or the doctrine of fluxions, intended as an Introduction to the Physico-Mathematical Sciences.* Etienne Bézout. Cambridge, 1824.

Farrar, Mrs. John. *Recollections of seventy years.* Boston: Tiknor and Fields, 1865.

Ferris, Timothy. *The Science of Liberty: Democracy, Reason, and the Laws of Nature.* New York: Harper, 2010.

Fisher, Alexander Metcalf. "On Maxima and Minima of Functions of two variable quantities." *American Journal of Science and Arts* 5 (1822):82-93.

Fiske, Thomas S. "Mathematical Progress in America." *Bulletin of the American Mathematical Society* 11 (1905):238-247.

Galuzzi, Massimo. "Review of *Mécanique analytique*," *Isis* 89 (1998):140-141.

Gillispie, Charles Coulston, ed. *The Dictionary of Scientific Biography*. New York: Scribner, 1981.

Good, H.G. *A History of Western Education*. New York: The Macmillan Company, 1947.

Greene John C. *American Science in the Age of Jefferson*. Ames: The Iowa State University Press, 1984.

Guicciardini, Niccoló. *The Development of Newtonian Calculus in Britain 1700-1800*. Cambridge: Cambridge University Press, 1989.

Guralnick, Stanley M. *Science and the Ante-Bellum American College*. Philadelphia: The American Philosophical Society, 1975.

Hall, A. Rupert. *Philosophers at War: The Quarrel Between Newton and Leibniz*. Cambridge: Cambridge University Press, 1980.

Heath, Thomas. *A History of Greek Mathematics*. New York: Dover Publications, Inc., 1981.

Hickey, Donald R. *The War of 1812*. Urbana and Chicago: University of Illinois Press, 1989.

Hindle, Brooke. "John Farrar." *Dictionary of Scientific Biography*, vol. 4, pp. 546-547.

——. *David Rittenhouse*. Princeton: Princeton University Press, 1964.

——. *The Pursuit of Science in Revolutionary America, 1735-1789*. Chapel Hill: University of North Carolina Press, 1956.

Historical Register of Harvard University, 1636-1936. Cambridge: Harvard University Press, 1937.

Hogan, Edward R. "George Baron and the *Mathematical Correspondent*." *Historia Mathematica* 3 (1976):403-415.

———. "Robert Adrain: American Mathematician." *Historia Mathematica* 4 (1977):157-172.

———. "The Mathematical Miscellany (1836-1839)." *Historia Mathematica* 12 (1985):245-257.

———. "Theodore Strong and Ante-Bellum American Mathematics." *Historia Mathematica* 8 (1981):439-455.

———. *Of the Human Heart: A Biography of Benjamin Peirce.* Bethlehem: Lehigh University Press, 2008.

Hoyt, Epaphras. "Astronomical observations made near the center of the village of Deerfield, Massachusetts." *Memoirs of the American Academy of Arts and Sciences* 3 (1815):305-307.

Jefferson, Thomas. *Notes on the State of Virginia.* Printed in Merrill D. Peterson, ed. *The Portable Thomas Jefferson.* New York: Penguin Books, 1975.

Jesseph, Douglass M. *Squaring the Circle.* Chicago: University of Chicago Press, 1999.

Kaplan, Lawrence S. *Jefferson and France: An Essay on Politics and Political Ideas.* New Haven: Yale University Press, 1967.

Karpinski, Louis. *Bibliography of Mathematical Works Printed in America Through 1850.* New York: Arno Press, 1980.

Katz, Victor J. *A History of Mathematics.* Reading MA: Addison-Wesley, 1998.

Kent, Deborah. "The Mathematical Miscellany and The Cambridge Miscellany of Mathematics: Closely connected attempts to introduce research-level mathematics in America, 1836-1843." *Historia Mathematica* 35 (2008): 102-122.

Kevles, Daniel. *The Physicists: a history of a scientific community in modern America.* Cambridge, MA: Harvard University Press, 1971.

Kline, Morris. *Mathematical Thought from Ancient to Modern Times.* Oxford: Oxford University Press, 1972.

Kuhn, Thomas S. *The Structure of Scientific Revolutions*, 3rd edition. Chicago: University of Chicago Press, 1996.

Lankford, John. *American Astronomy: community, careers, and power, 1859-1940*. Chicago: University of Chicago Press, 1997.

Mansfield, Jared. *Essays, Mathematical and Physical*. New Haven, CT: William W. Morse, 1802.

Mapp, Alf J. *Thomas Jefferson: Passionate Pilgrim*. New York: Madison Books, 1991.

Morrel, Jack and Thackray, Arnold, eds. *Gentlemen of Science: Early Correspondence of the British Association for the Advancement of Science*. London: Royal Historical Society, 1984.

Newcomb, Simon. "Exact Science in America." *North American Review* 119 (1874):286-308.

Newton, H.A. "Benjamin Peirce." *Proceedings of the American Academy of Arts and Sciences*, n.s. 8 (1881):443-454.

Nulty, Eugenius. "On the Motions of Solids on Surfaces." *Transactions of the American Philosophical Society*. 2 (1825):466-477.

Nulty, Owen. "Investigation of a Theorem, proposed by Dr. Rittenhouse, respecting the summation of the several Powers of the Sines; with its Application to the Problem of a Pendulum vibrating in circular Arcs." *Transactions of the American Philosophical Society*. 1(new series) (1818):395-400.

Nye, Russel Blaine. *The Cultural Life of the New Nation*. New York: Harper and Row, 1960.

"Obituary of Professor Fisher," *American Journal of Science and Arts* 5 (1822):367-376.

"Obituary of Robert Adrain." *U.S. Magazine and Democratic Review* 14 (1844):646-652.

O' Donnel, Terence. *The History of Life Insurance in Its Formative Years*. Chicago: American Conservation Company, 1936.

Oleson, Alexandra Brown, Sanborn C., eds. *The Pursuit of Knowledge in the Early American Republic: American Scientific and Learned Societies from Colonial Times to the Civil War.* Baltimore: Johns Hopkins University Press, 1976.

Palfrey, John G. "John Farrar." *The Christian Examiner and Religious Miscellany.* 50 (1853):121-136.

Parshall, Karen Hunger and Rowe, David E. *The Emergence of the American Mathematical Research Community, 1876-1900: J.J. Sylvester, Felix Klein, and E.H. Moore.* Providence, RI: The American Mathematical Society, 1994.

Patterson, Robert. "An improved Method of projecting and measuring plane Angles." *Transactions of the American Philosophical Society* 6 (1809):29-31.

Pickering, John. "Mr. Pickering's Eulogy on Dr. Bowditch, President of the Academy." *Memoirs of the American Academy of Arts and Sciences* 2 (1846).

Portolano, Marlana. "John Quincy Adams' Rhetorical Crusade for Astronomy." *Isis* 91 (2000):480-503.

Price, Derek de Solla. "Toward a Model for Science Indicators." in Yehuda Elkana, et. al., eds. *Toward a Metric of Science: The Advent of Science Indicators.* New York: John Wiley & Sons, 1978, pp. 69-96.

Proclus (pseudonym). "On the Principles of Motion, and their use in the higher branches of Mathematics." *American Journal of Science and Arts* 14 (1828):297-302.

Pycior, Helena. "British Synthetic vs. French Analytical Styles of Algebra in the Early American Republic." in David Rowe and John Mcleary, eds. *History of Modern Mathematics*, vol. 1.Boston: Academic Press, 1989, pp. 125-154.

———. "Benjamin Peirce's *Linear Associative Algebra*." *ISIS* 70 (1979):537-551.

Quinby, A.B. "On Crank Motion." *American Journal of Science and Arts* 7 (1824):316-322.

———. "Problem to determine the position of the Crank when the tendency of the power to produce the rotation is maximum." *American Journal of Science and Arts* 11 (1826):338-339.

Reid-Maroney, Nina. *Philadelphia's Enlightenment, 1740-1800*. Westport, Connecticut: Greenwood Press, 2001.

Reingold, Nathan, ed. *The Sciences in the American Context: New Perspectives*. Washington, D.C.: Smithsonian Institution Press, 1979.

———. "Nathaniel Bowditch." *Dictionary of Scientific Biography*, vol. 2, pp. 368-369.

———. *Science in Nineteenth-Century America*. New York: Hill and Wang, 1964.

"Robert Adrain." *U.S. Magazine and Democratic Review* 14 (1844):646-652.

Rosenstein, George M. "The Best Method. American Calculus Textbooks of the Nineteenth Century." in Duren, Peter, ed. *A Century of Mathematics in America*, vol. 3. Providence, RI: American Mathematical Society, 1988, pp. 77-109.

Rudolph, Frederick. *Curriculum: A History of the American Undergraduate Course of Study Since 1636*. San Francisco: Jossey-Bass, Inc., 1977.

Schneider, Ivo. "Laplace and Thereafter: The Status of Probability Calculus in the Nineteenth Century." in Krüger, Lorenz; Daston, Lorraine; and Heidelberger, Michael, eds. *The Probabilistic Revolution*, vol. 1.Cambridge, MA: The MIT Press, 1987, pp. 191-214.

Servos, John. *Physical Chemistry from Ostwald to Pauling: the making of a science in America*. Princeton: Princeton University Press.

Simons Lao G. "The Influence of French Mathematicians at the end of the Eighteenth Century upon the Teaching of Mathematics in American Colleges." *ISIS* 15 (1931):104-123.

Sinclair, Bruce. "Americans Abroad: Science and Cultural Nationalism in the Early Nineteenth Century." in Nathan Reingold, ed. *The Sciences in the American Context: New Perspectives.* Washington, D.C.: Smithsonian Institution Press, 1979, pp. 35-53.

Smith, David Eugene and Ginsburg, Jekuthiel. *A History of Mathematics in America Before 1900.* New York: Arno Press, 1980.

Smith, David Eugene. "Early American Mathematical Periodicals." *Scripta Mathematica* 1 (1932-1933): 277-285.

Smith, James Morton, ed. *The Republic of Letters: The Correspondence between Thomas Jefferson and James Madison 1776-1826.* New York: W.W. Norton and Co., 1995.

Stearns, Raymond Phineas. *Science in the British Colonies of America.* Urbana: University of Illinois Press, 1970.

Stigler, Stephen. "Mathematical Statistics in the Early States." in Duren, Peter, ed. *A Century of Mathematics in America*, vol. 3. Providence, RI: American Mathematical Society, 1988, pp. 537-564.

———. *The History of Statistics.* Cambridge: Cambridge University Press, 1990.

Strong, Theodore. "An improved Method of obtaining the Formulae for the Sines and Cosines of The Sum and Difference of two Arcs." *American Journal of Science and Arts* 1 (1819):424-427.

———. "Mathematical Problems, with geometrical constructions and demonstrations." *American Journal of Science and Arts* 2 (1820):54-64.

———. "Solution of a Problem in Fluxions." *American Journal of Science and Arts* 16 (1829):283-287.

———. "Solution of a Problem in Fluxions." *American Journal of Science and Arts* 17 (1830):69-73.

Struik, Dirk J. "Robert Adrain." *Dictionary of Scientific Biography,* vol. 1, pp. 65-66.

———. *Yankee Science in the Making: Science and Engineering in New England from Colonial Times to the Civil War.* New York: Dover Publications, 1991.

Timmons, Todd. "A Prosopographical Analysis of the Early American Mathematics Community." *Historia Mathematica* 31 (2004): 429-454.

de Tocqueville, Alexis. *Democracy in America.* New York: Harper and Row, 1988.

Wallace. "Professor Wallace, in reply to the Remarks of B. upon his paper on Algebraic Series." *American Journal of Science and Arts* 9 (1825):98-103.

Weiss, Paul. "Charles Sanders Peirce." *Dictionary of American Biography,* vol. 14, pp. 482-487.

West, Benjamin. "On the Extraction of Roots." *Memoirs of the American Academy of Arts and Sciences* 1 (1785):165-172.

Wheeler, A.D. "An Easy Solution of a Diophantine Problem." *American Journal of Science and Arts* 20 (1831):295-297.

Wilder, C. "Algebraic Solutions." *American Journal of Science and Arts* 16 (1829):271-282.

Wilkes, M.V. "Herschel, Peacock, Babbage and the Development of the Cambridge Curriculum." *Notes Rec. R. Soc. Lond* 44 (1990):205-219.

Winthrop, James. "Geometrical Methods of finding any required Series of Mean Proportionals between given Extremes." *Memoirs of the American Academy of Arts and Sciences* 2 (1793):9-13.

———. "A Rule for Trisecting Angles geometrically." *Memoirs of the American Academy of Arts and Sciences* 2 (1793):14-17.

Wright, Elizur. "A Theory of Fluxions." *American Journal of Science and Arts* 14 (1828):330-350.

———. "A Discourse on the different views that have been taken of the Theory of Fluxions." *American Journal of Science and Arts* 16 (1829):53-59.

Young, Alexander. *A Discourse on the Life and Character of the Hon. Nathaniel Bowditch, LL.D, F.R.S.* Boston: Charles Little and James Brown, 1838.

Zitarelli, David E. "The Bicentennial of American Mathematics Journals." *The College Mathematics Journal* 36 (2005): 2-15.

Appendix A

Mathematical Content of American Journals, 1771—1834

AAAS	*Memoirs of the American Academy of Arts and Sciences*
AJSA	*American Journal of Science and Arts*
ANALYST	*The Analyst*
APS	*Transactions of the American Philosophical Society*
CAAS	*Connecticut Academy of Arts and Sciences*
Correspondent	*The Mathematical Correspondent*
DIARY	*The Mathematical Diary*
Miscellany	*The Mathematical Miscellany*
MUSEUM	*The Analyst, or Mathematical Museum*

Table A.1: Journal Abbreviations

P	Pure Mathematics
M	Mixed Mathematics
O	Other non-mathematical articles by men connected to mathematics
R	Reviews and mathematical exposition

Table A.2: Other Abbreviations

	APS			AAAS			AJSA			
Year	P	M	O	P	M	O	P	M	O	R
1771		42.5								
1785				1.4	36.0	0.4				
1786		11.3	2.5							
1793	0.6	2.6	6.0	10.0	8.5	2.0				
1799	0.6	14.5	1.1							
1802		17.6								
1804				1.8	26.0					
1809	1.5	12.0	0.5	3.3	15.7	4.3				
1815					62.9	34.6				
1818	4.9	14.4	5.6		33.1	5.5			15.6	
1819							0.5			
1820					17.5	5.0	4.7			
1821									0.8	
1822							3.3	3.0	1.0	4.0
1823							0.8		3.8	4.8
1824							1.0	7.9	1.3	3.1
1825		41.9					1.5	8.0	2.8	4.2
1826							1.4	3.9	1.3	0.1
1827							1.0	4.3	3.5	4.8
1828							2.6	2.1	2.6	1.5
1829							2.2	4.5	1.0	1.0
1830		14.7					3.8	1.2	1.2	0.2
1831							1.2	4.1	0.5	
1832							2.0	2.2		
1833					8.5		2.1	5.0	0.6	
1834		3.2	2.2				1.7	11.7		1.7

Table A.3: Percent of Pages Devoted to Mathematics by Volume

Appendix B

Contributions to American Mathematics Journals

B.1 *American Journal of Science and Arts,* 1818—1834

Name	Residence	Profession
Abbott, Jacob	Massachusetts	Professor at Amherst
Adams, Jasper	Rhode Island	Professor at Brown
Allen, Rev. William	Maine	President of Bowdoin
Allen, Zachariah	Rhode Island	Inventor
Blake, Eli W.	Connecticut	Firearms business
Burritt, E.H.	Connecticut	Blacksmith
Cleaveland, Parker	Maine	Professor at Bowdoin

Name	Residence	Profession
Davies, Charles	New York	Professor at West Point
Dean, James	Vermont	Professor at Vermont
DeWitt, Simeon	New York	Surveyor
Doolittle, Isaac		
Fisher, Alexander M.	Connecticut	Professor at Yale
Fourier, Baron	France	
French, Stiles		
Gould, David		
Gregory, Olinthus	England	
Hazard, Rowland G.	Rhode Island	Manufacturer
Hitchcock, Rev. Edward	Massachusetts	Professor at Amherst
Johnson, Edwin	Vermont	Civil Engineer, Professor at Norwich
Kendall, Thomas		

Name	Residence	Profession
Lyon, Lucius	Michigan	Surveyor, Civil Engineer
Maury, M.F.	Virginia	U.S. Navy
Olmsted, Denison	Connecticut	Professor at North Carolina & Yale
Paucalis, Felix		
Quinby, A.B.	New York	
Renwick, James	New York	Professor at Columbia
Rodriguez, P.J.		
Strong, Theodore	New Jersey	Professor at Hamilton & Rutgers
Thomson, J.	Tennessee	Civil Engineer
Tiarks, J.L.		Astronomer
Twining, Alexander	Connecticut	Civil Engineer
Wallace, J.	South Carolina	
Wheeler, A.D.	Massachusetts	Teacher at Latin School
Wilder, C.	New Orleans	

Name	Residence	Profession
Wilkie, Rev. Daniel		Clergyman, Actuary
Wright, Elizer	Connecticut Ohio Massachusetts	Actuary, Professor at Western Reserve

B.2 *Transactions of the American Philosophical Society,* 1771—1834

Name	Residence	Profession
Adrain, Robert	New York, New Jersey, Pennsylvania	Professor at Columbia, Rutgers, & Pennsylvania
Anderson, Henry J.	New York	Professor at Columbia
Austin, James	Pennsylvania	Lawyer, Politician
Biddle, Owen	Pennsylvania	Clock-maker
Bradley, John		
Clay, Joseph	Pennsylvania	Clergyman, Lawyer
de Ferrer, Jose J.	Spain	Astronomer
Ellicott, Andrew	Pennsylvania, New York	Surveyor, Professor at West Point
Ewing, Rev. John	Pennsylvania	Provost at Pennsylvania
Gunmere, John	New Jersey	Teacher at boarding schools

Name	Residence	Profession
Hassler, Ferdinand	New York	Professor at West Point, U.S. Coast Survey
Joslin, Benjamin F.	New York	Professor at Union
Lambert, William	Washington, D.C.	
Mansfield, Jared	New York	Professor at West Point
Nancarrow, Andrew		
Nulty, Eugenius	Pennsylvania	Private tutor
Nulty, Owen	Pennsylvania	Professor at Dickinson
Partridge, Alden	New York, Vermont	Professor at West Point Army Corps of Engineers
Patterson, Robert	Pennsylvania	Professor and Provost at Pennsylvania
Rittenhouse, David	Pennsylvania	Surveyor, Director of the Mint, Instrument-maker

Name	Residence	Profession
Smith, Rev. William	Pennsylvania	Clergyman, Teacher at College of Philadelphia
West, Benjamin	Rhode Island	Almanac-maker, Professor at Brown
Williams, Samuel	Massachusetts	Professor at Harvard
Williamson, Hugh	Pennsylvania	Physician, Professor at College of Philadelphia

B.3 *Memoirs of the American Academy of Arts and Sciences*, 1785—1833

Name	Residence	Profession
Baron, George	New York	Professor at West Point
Bond, W. Cranch	Massachusetts	Astronomer
Bowditch, Nathaniel	Massachusetts	Insurance executive, Actuary
Bowdoin, James	Massachusetts	Diplomat, Merchant
Brown, Joseph	Rhode Island	Professor at Brown
Cleaveland, Parker	Maine	Professor at Bowdoin
Crosswell, William		Teacher of Navigation
Cutler, Rev. Manesseh	Massachusetts	Clergyman, Botanist
Dean, James	Vermont	Professor at Vermont
Dearborn, Benjamin	Massachusetts	

Name	Residence	Profession
Dewey, Chester	Massachusetts	Professor at Williams
Farrar, John	Massachusetts	Professor at Harvard
Fisher, Alexander M.	Connecticut	Professor at Yale
Folger, Walter	Massachusetts	Instrument-maker, Attorney
Gannett, Caleb	Massachusetts	
Hoyt, Epaphras	Massachusetts	Surveyor
Mercator (pseudonym?)		
Nichols, Rev. Ichabod	Maine	Clergyman, Tutor at Harvard
Paine, Robert T.	Massachusetts	Attorney, Manufacturer
Parsons, Theophilus	Massachusetts	Jurist
Payson, Rev. Phillips	Massachusetts	Clergyman
Peters, Joseph		

Name	Residence	Profession
Pope, Joseph		
Schubert, F.T.	Russia	
Webber, Samuel	Massachusetts	Professor, President at Harvard
West, Benjamin	Rhode Island	
Willard, Joseph	Massachusetts	President at Harvard
Williams, Samuel	Massachusetts	Professor at Harvard
Winthrop, James	Massachusetts	Librarian at Harvard
Wright, Elizur	Connecticut, Ohio, Massachusetts	Actuary, Professor at Western Reserve

Appendix C

American Mathematicians Published in Major American Science Journals, 1771—1834

Abbott, Jacob (1803-1879)

- Graduated from Bowdoin College in 1820.
- Professor of mathematics and natural philosophy at Amherst (1825-1829).
- (P)AJSA, 1829, *15*(No. 2):368.

Adams, Jasper (1793-1841)

- Graduated from Brown in 1815.
- Professor of mathematics and natural philosophy at Brown (1819-1824).
- (R)AJSA, 1822, *5*:304-335.
- (R)AJSA, 1823, *6*:283-302.
- (O)AJSA, 1824, *8*:121-130.

Adrain, Robert (1775-1843)

- Born in Ireland. Immigrated to the U.S. in 1798.
- Professor of mathematics at Queen's College, now Rutgers (1809-1813).
- Professor of mathematics and natural philosophy at Columbia College (1813-1826).
- Professor of mathematics at the University of Pennsylvania (1827-1834).
- Founded and/or edited several mathematics journals, including *The Mathematical Correspondent* (editor), *The Analyst, or Mathematical Museum, The Analyst* and *The Mathematical Diary*.
- Published in *Correspondent, Museum, Analyst, Diary*.
- Prepared the American edition to Hutton's *Course of Mathematics*.
- Published solutions to the error distribution problem a year before Gauss.
- (M)APS, 1818, *1*(n.s.):119-136.
- (M)APS, 1818, *1*(n.s.):353-366.

Allen, Rev. William (1784-1868)

- Graduated from Harvard in 1802.
- President of Dartmouth (1817-1820),
- President of Bowdoin College (1820-1839).
- (P)AJSA, 1822, *4*:343-356.

Allen, Zachariah (1795-1882)

- Graduated from Brown in 1815.
- Rhode Island inventor.
- (M)AJSA, 1830, *17*(No. 2):338-344.

Anderson, Henry James (1799-1875)

- Graduated from Columbia in 1818.
- Professor of Mathematics and Astronomy at Columbia (1825-1850).
- (M)APS, 1830, *3*(n.s.):315-382.

Austin, James (1784-1870)

- Graduated from Harvard in 1802.
- Lawyer and Attorney General of Massachusetts.
- (P)APS, 1818, *1*(n.s.):181-186.

Baron, George (1769-?)

- Englishman who moved back to England sometime after 1806.
- Acting Professor of Mathematics at West Point, January 1801 - February 1802. (Before West Point was officially founded in 1802).
- Founded *The Mathematical Correspondent* in 1804, the first mathematical journal in the United States.
- (P)AAAS, 1804, *2*(part 2):40-42.

Biddle, Owen
- Pennsylvania resident.
- (M)APS, 1771, *1*:89-96.

Blake, Eli W. (1795-1886)
- Nephew of Eli Whitney.
- Graduated from Yale in 1816.
- Founder and president of CAAS.
- Wrote *Original Solutions of Several Problems in Aerodynamics* (1882).
- Spent life in manufacturing, firearms business of his uncle Eli Whitney.
- (M)AJSA, 1824, *7*:86-102.
- (R)AJSA, 1827, *12*(No. 1):338-343.
- (R)AJSA, 1828, *13*(No. 1):75-76.
- (R)AJSA, 1828, *13*(No. 2):350-355.

Bond, William Cranch (1789-1859)
- First director of the Harvard Observatory.
- (M)AAAS, 1833, *1*(new series):79-83.
- (M)AAAS, 1833, *1*(new series):84-90.

Bowditch, Nathaniel (1773-1838)
- Self-taught mathematician, navigator, surveyor and scientist.
- Published *New American Practical Navigator* in 1802.

- Served as president of AAAS from 1829-1838.
- Translated Laplace's *Mécanique Céleste* between 1815-1817. Not published until 1829-1839.
- Turned down several academic positions. Spent life working in insurance industry.
- (M)AAAS, 1804, *2*(part 2):1-11.
- (M)AAAS, 1809, *3*(part 1):1-17.
- (M)AAAS, 1809, *3*(part 1):18-22.
- (M)AAAS, 1809, *3*(part 1):23-32.
- (P)AAAS, 1809, *3*(part 1):33-37.
- (M)AAAS, 1815, *3*(part 2):213-236.
- (M)AAAS, 1815, *3*(part 2):255-304.
- (M)AAAS, 1815, *3*(part 2):313-325.
- (M)AAAS, 1815, *3*(part 2):337-343.
- (M)AAAS, 1815, *3*(part 2):413-436.
- (M)AAAS, 1815, *3*(part 2):437-438.
- (M)AAAS, 1818, *4*(part 1):28-29.
- (M)AAAS, 1818, *4*(part 1):30-49.
- (M)AAAS, 1818, *4*(part 1):50-56.
- (M)AAAS, 1818, *4*(part 1):57-61.
- (M)AAAS, 1818, *4*(part 1):62-73.
- (M)AAAS, 1818, *4*(part 1):74-75.

- (M)AAAS, 1818, *4*(part 1):110-119.
- (M)AAAS, 1820, *4*(part 2):295-305.
- (M)AAAS, 1820, *4*(part 2):306.
- (M)AAAS, 1820, *4*(part 2):307-308.
- (M)AAAS, 1820, *4*(part 2):317-318.
- (R)AJSA, 1825, *9*:293-303.

Bowdoin, James (1727-1790)

- Diplomat and merchant in Massachusetts.
- First president of the AAAS.
- (M)AAAS, 1785, *1*:208-233.

Bradley, John

- (M)APS, 1771, *1*:114-116.

Brewster, David

- (M)AJSA, 1833, *23*(No. 2):225-236.

Brown, Joseph

- Professor at Brown University.
- (M)AAAS, 1785, *1*:149-150.

Burritt, E.H.

- Connecticut blacksmith.
- (M)AJSA, 1834, *26*(No. 1):129-131.

Clay, Joseph

- Graduated from Princeton.
- Clergyman, lawyer and jurist from Philadelphia.
- (M)APS, 1802, *5*:312-319.

Cleaveland, Parker (1780-1858)

- Graduated from Harvard in 1799.
- Professor of Mathematics and Natural Philosophy at Bowdoin College (1805-1858).
- In spite of title, primarily known as a mineralogist.
- Published *An Elementary Treatise on Mineralogy and Geology* (1816).
- (O)AAAS, 1809, *3*(part 1):119-121.
- (O)AAAS, 1809, *3*(part 1):153-158.
- (M)AAAS, 1815, *3*(part 2):247-248.
- (M)AAAS, 1818, *4*(part 1):120-128.
- (O)AJSA, 1823, *6*:162.
- (M)AJSA, 1826, *10*:129.

Croswell, William

- Teacher of navigation.
- (P)AAAS, 1793, *2*:18-19.
- (P)AAAS, 1809, *3*(part 1):38-39.

Cutler, Rev. Manasseli (1742-1823)

- Graduated from Yale in 1765.
- Massachusetts clergyman and botanist.
- (M)AAAS, 1785, *1*:128.
- (M)AAAS, 1785, *1*:162-164.

Davies, Charles (1798-1876)

- Graduated from West Point, 1815.
- Professor of mathematics at West Point, May 1823 – May 1837.
- (P)AJSA, 1823, *6*:280-282.

Dean, James (1776-1849)

- Graduated from Dartmouth in 1800.
- Tutor at University of Vermont (1807-1809).
- Professor of mathematics and natural philosophy at the University of Vermont (1809-1814 and 1821-1824).
- Professor of mathematics and natural philosophy at the Dartmouth (1814-1821).
- (M)AAAS, 1815, *3*(part 2):241-245.
- (M)AAAS, 1815, *3*(part 2):249-251.
- (M)AAAS, 1815, *3*(part 2):329-332.
- (M)AAAS, 1815, *3*(part 2):344-345.
- (O)AJSA, 1823, *6*:322-325.

Dearborn, Benjamin (1755-1838)

- Massachusetts inventor.
- (M)AAAS, 1804, *2*(part 2):23-24.

Dewey, Chester (1784-1867)

- Graduated from Williams College in 1806.
- Professor of mathematics and natural philosophy at Williams College (1810-1827).
- Professor of chemistry and natural philosophy at University of Rochester (1850-1860).
- (O)AAAS, 1820, *4*(part 2):387-392.

DeWitt, Simeon

- New York surveyor.
- (P)AJSA, 1833, *24*(No. 2):369.

Doolittle, Isaac

- (M)AJSA, 1822, *4*:102-123.
- (M)AJSA, 1824, *7*:286-315.
- (M)AJSA, 1828, *14*(No. 1):60-62.

Ellicott, Andrew (1754-1820)

- Pupil of Robert Patterson
- Professor of Mathematics at West Point, September 1813 – August 1820.
- Employed by government as surveyor. Surveyed site of Washington, D. C.

- (O)APS, 1793, *3*:62-63.
- (M)APS, 1793, *3*:116-118.
- (M)APS, 1799, *4*:32-50.
- (M)APS, 1799, *4*:51-66.
- (M)APS, 1799, *4*:67-68.
- (M)APS, 1799, *4*:224-230.
- (M)APS, 1799, *4*:231.
- (M)APS, 1799, *4*:447-451.
- (M)APS, 1802, *5*:162-202.
- (M)APS, 1802, *5*:203-311.
- (M)APS, 1809, *6*:26-27.
- (O)APS, 1809, *6*:28.
- (M)APS, 1809, *6*:59.
- (M)APS, 1809, *6*:61-68.
- (M)APS, 1809, *6*:113-118.
- (M)APS, 1818, *1*(n.s.):93-102.

Ewing, Rev. John (1732-1802)

- Graduated from Princeton in 1754.
- Provost of University of Pennsylvania (1779-1802).
- (M)APS, 1771, *1*:5-7.
- (M)APS, 1771, *1*:42-88.

- (M)APS, 1771, *1*:Appendix(21-26).

Farrar, John (1779-1853)

- Graduated from Harvard in 1803.
- Hollis Professor of Mathematics and Natural Philosophy, Harvard (1807-1836).
- Translated popular series of French textbooks known as the Cambridge Mathematics Series and Cambridge Natural Philosophy Series.
- (M)AAAS, 1815, *3*(part 2):308-312.
- (O)AAAS, 1815, *3*(part 2):361-398.
- (O)AAAS, 1815, *3*(part 2):399-412.
- (O)AAAS, 1818, *4*(part 1):92-97.
- (O)AAAS, 1818, *4*(part 1):98-107.

Fisher, Alexander M. (1794-1822)

- Graduated from Yale in 1813.
- Professor of Mathematics and Natural Philosophy at Yale, 1817-1822.
- (O)AJSA, 1818, *1*(No. 1):9-34.
- (O)AJSA, 1818, *1*(No. 2):176-199.
- (M)AAAS, 1820, *4*(part 2):309-316.
- (O)AJSA, 1821, *3*:326.
- (P)AJSA, 1822, *5*:82-93.

Folger, Walter (1765-1849)

- Massachusetts instrument-maker and attorney.
- Published in *Correspondent*.
- (M)AAAS, 1815, *3*(part 2):252-254.

French, Stiles

- (P)AJSA, 1830, *17*(No. 1):74-80.

Gannett, Caleb

- Massachusetts resident.
- (M)AAAS, 1785, *1*:146-148.

Gould, David

- (P)AJSA, 1831, *19*(No. 1):50-51.

Gummere, John (1784-1845)

- Self-taught mathematician.
- Teacher at several New Jersey boarding schools.
- (M)APS, 1830, *3*(n.s.):467-470.

Hassler, Ferdinand R. (1770-1843)

- Born in Switzerland, immigrated to the U.S. in 1805.
- Acting Professor of Mathematics, February 14, 1807–December 31, 1809 at West Point.
- Briefly served as professor of mathematics and natural philosophy at Union College (1810-1811).
- First superintendent of the United States Coast Survey.

- (M)APS, 1818, *1*(n.s.):210-227.
- (M)APS, 1825, *2*(n.s.):232-420.

Hazard, Rowland

- Rhode Island manufacturer.
- (P)AJSA, 1832, *21*(No. 2):314-315.

Hitchcock, Rev. E. (1793-1864)

- Professor of chemistry and natural philosophy at Amherst (1825-1845).
- Massachusetts state geologist (1830-1844).
- President of Amherst (1845-1854).
- (M)AJSA, 1825, *9*:107-118.
- (M)AJSA, 1834, *25*(No. 2):354-362.

Hoyt, Epaphras (1765-1850)

- Massachusetts resident.
- (M)AAAS, 1815, *3*(part 2):305-307.

Johnson, Edwin F.

- Civil engineer, professor at Norwich University, Vermont.
- (M)AJSA, 1831, *19*(No. 1):131-140.
- (P)AJSA, 1832, *21*(No. 2):280-283.

Joslin, Benjamin F., M.D.

- Professor at Union College, New York.

- (O)APS, 1834, *4*(n.s.):340-350.

Kendall, Thomas

- From New Lebanon, N.Y.
- (M)AJSA, 1831, *19*(No. 2):337-338.

Lambert, William

- Washington, D.C. resident.
- Also published using the pseudonym Mary Bond.
- (M)APS, 1818, *1*(n.s.):103-118.

Littrow, J.J.

- (M)AJSA, 1833, *24*(No. 2):346-348.

Lyon, Lucius (1800-1851)

- Surveyor and civil engineer.
- (M)AJSA, 1828, *14*(No. 2):268-275.

Mansfield, Jared (1759-1830)

- New York resident.
- Graduated from Yale in 1777.
- Wrote *Essays, Mathematical and Physical* (1801). Considered the first book of original mathematics by a native American.
- Acting Professor of Mathematics, May 3, 1802-November 14, 1803, at West Point.
- Published several papers in *CAAS*.
- (P)APS, 1818, *1*(n.s.):200-209.

Maury, Matthew F. (1806-1873)

- From Virginia
- First Superintendent of the Naval Observatory.
- (M)AJSA, 1834, *26*(No. 1):63-64.

Nancarrow, John

- (M)APS, 1799, *4*:348-361.

Nichols, Rev. Ichabod (1784-1859)

- Graduated from Harvard in 1802.
- Tutor of mathematics at Harvard (1803-1809).
- Maine resident.
- (M)AAAS, 1815, *3*(part 2):246.

Nulty, Eugenius

- (M)APS, 1825, *2*(n.s.):466-477.

Nulty, Owen

- Professor of mathematics at Dickenson College in Pennsylvania.
- (M)APS, 1818, *1*(n.s.):395-400.

Olmsted, Denison (1791-1859)

- Graduated from Yale in 1813.
- Professor of chemistry, mineralogy and geology at University of North Carolina (1817-1825).

- Professor of mathematics and natural philosophy at Yale (1825-1836).
- (O)AJSA, 1821, *3*:100-101.
- (O)AJSA, 1822, *5*:257-264.
- (O)AJSA, 1825, *9*:5-15.
- (O)AJSA, 1826, *11*(No. 2):349-358.
- (R)AJSA, 1827, *12*(No. 2):359-363.
- (O)AJSA, 1828, *14*(No. 2):230-250.
- (O)AJSA, 1829, *16*(No. 1):70-77.
- (O)AJSA, 1830, *18*(No. 1):1-10.
- (O)AJSA, 1831, *20*(No. 2):373-376.
- (M)AJSA, 1834, *25*(No. 2):363-410.
- (M)AJSA, 1834, *26*(No. 1):132-173.

Paine, Robert T.
- Massachusetts attorney and manufacturer.
- (M)AAAS, 1833, *1*(new series):45-69.
- (M)AAAS, 1833, *1*(new series):70-72.
- (M)AAAS, 1833, *1*(new series):338-452.

Parsons, Theophilus (1750-1813)
- Graduated from Harvard in 1769.
- Massachusetts jurist.
- (M)AAAS, 1804, *2*(part 2):12-19.

Partridge, Alden (1785-1854)

- Graduated from West Point in 1806.
- Acting Professor of Mathematics at West Point, December 1809 – April 1813.
- Professor of Mathematics at West Point, April 1813 – September 1813.
- Corps of Engineers.
- (M)APS, 1818, *1*(n.s.):147-150.

Patterson, Robert (1743-1824)

- Born in Ireland, immigrated to America in 1768.
- Professor of mathematics at University of Pennsylvania (1779-1814).
- Vice-Provost of University of Pennsylvania (1810-1813).
- Director of the Mint for US (1805-1824).
- President of APS (1819-1824).
- Published in the *Analyst*.
- *Lectures on Select Subjects in Mechanics* (1806).
- *Astronomy Explained Upon Sir Isaac Newton's Principles* (1806,1809).
- *Newtonian System of Philosophy* (1808).
- *A Treatise on Practical Arithmetic* (1818).
- (M)APS, 1786, *2*:251-259.
- (O)APS, 1793, *3*:13-16.

- (O)APS, 1793, *3*:139-143.
- (O)APS, 1793, *3*:321-323.
- (M)APS, 1799, *4*:154-161.
- (P)APS, 1809, *6*:29-31.
- (M)APS, 1809, *6*:59.
- (M)APS, 1818, *1*(n.s.):325-332.
- (M)APS, 1818, *1*(n.s.):333-339.
- (O)APS, 1818, *1*(n.s.):367-370.
- (O)APS, 1818, *1*(n.s.):427-429.

Paucalis, Felix (M.D.)

- (M)AJSA, 1826, *11*(No. 2):339-348.

Payson, Rev. Phillips (1736-1801)

- Graduated from Harvard in 1754.
- Massachusetts clergyman.
- (M)AAAS, 1785, *1*:124-127.

Peters, Joseph

- (M)AAAS, 1785, *1*:143-145.

Pope, Joseph

- (M)AAAS, 1804, *2*(part 2):43-45.

Quinby, A. B.

- New York resident.

- (M)AJSA, 1824, *7*:316-322.
- (M)AJSA, 1825, *9*:304-312.
- (M)AJSA, 1825, *9*:313-315.
- (P)AJSA, 1825, *9*:316.
- (M)AJSA, 1825, *9*:317-323.
- (M)AJSA, 1826, *11*(No. 2):333-337.
- (M)AJSA, 1826, *11*(No. 2):338.
- (M)AJSA, 1827, *12*(No. 1):128-131.
- (R)AJSA, 1827, *12*(No. 2):344-345.
- (M)AJSA, 1827, *12*(No. 2):346-358.
- (R)AJSA, 1828, *13*(No. 1):73-74.
- (R)AJSA, 1828, *13*(No. 1):356-357.

Renwick, James (1790-1863)

- Graduated from Columbia in 1807.
- Professor of natural and experimental philosophy and chemistry at Columbia (1813-1853).
- (M)AJSA, 1822, *5*:143.

Rittenhouse, David (1732-1796)

- Succeeded Benjamin Franklin as president of APS.
- Surveyor, instrument-maker.
- First director of U.S. Mint.

- Professor of astronomy at the University of Pennsylvania (1779-1782).
- (M)APS, 1771, *1*:1-3.
- (M)APS, 1771, *1*:4.
- (M)APS, 1771, *1*:Appendix(37-44).
- (M)APS, 1771, *1*:Appendix(47-49).
- (M)APS, 1786, *2*:37-41.
- (O)APS, 1786, *2*:173-176.
- (O)APS, 1786, *2*:178-180.
- (M)APS, 1786, *2*:181-182.
- (M)APS, 1786, *2*:195.
- (M)APS, 1786, *2*:201-205.
- (M)APS, 1786, *2*:260-262.
- (O)APS, 1793, *3*:119-121.
- (O)APS, 1793, *3*:122-124.
- (M)APS, 1793, *3*:150-154.
- (P)APS, 1793, *3*:155-156.
- (O)APS, 1793, *3*:261.
- (M)APS, 1799, *4*:21-25.
- (O)APS, 1799, *4*:26-28.
- (O)APS, 1799, *4*:29-31.
- (P)APS, 1799, *4*:69-71.

Rodriguez, P.J.

- (M)AJSA, 1829, *16*(No. 1):94-98.

Smith, Rev. William (1727-1803)

- Graduated from the University at Aberdeen, Scotland in 1747.
- Pennsylvania clergyman.
- Professor at seminary in Philadelphia (later the University of Pennsylvania).
- (M)APS, 1771, *1*:8-41.
- (M)APS, 1771, *1*:105-113.
- (M)APS, 1771, *1*:Appendix(5-11).
- (M)APS, 1771, *1*:Appendix(50-53).
- (M)APS, 1771, *1*:Appendix(54-?).

Strong, Theodore (1790-1869)

- Graduated from Yale in 1812.
- Professor of mathematics and natural philosophy at Hamilton College (1816-1827).
- Professor of mathematics and natural philosophy at Rutgers (1827-1861).
- Vice-president of Rutgers (1839-1863).
- Published in *CAAS, Miscellany, Diary*.
- Wrote one algebra text (1859) and one calculus text (1869).
- One of 50 original incorporators of the National Academy of Science.

- Taught George William Hill, who took advantage of Strong's extensive library of mathematical classics.
- (P)AJSA, 1819, *1*(No. 4):424-427.
- (P)AJSA, 1820, *2*(No. 1):54-64.
- (P)AJSA, 1820, *2*(No. 2):266-280.
- (P)AJSA, 1827, *12*(No. 1):132-135.
- (P)AJSA, 1829, *16*(No. 2):283-287.
- (P)AJSA, 1830, *17*(No. 1):69-73.
- (P)AJSA, 1830, *17*(No. 2):329-333.
- (P)AJSA, 1830, *18*(No. 1):67-69.
- (M)AJSA, 1830, *18*(No. 1):70-71.
- (M)AJSA, 1831, *19*(No. 1):46-49.
- (M)AJSA, 1831, *20*(No. 1):65-73.
- (M)AJSA, 1831, *20*(No. 2):291-294.
- (M)AJSA, 1832, *21*(No. 1):66-68.
- (M)AJSA, 1832, *21*(No. 2):334-341.
- (M)AJSA, 1832, *22*(No. 1):132-135.
- (M)AJSA, 1832, *22*(No. 2):343-345.
- (M)AJSA, 1833, *24*(No. 1):40-45.
- (M)AJSA, 1834, *25*(No. 2):281-289.
- (M)AJSA, 1834, *26*(No. 1):44-53.
- (M)AJSA, 1834, *26*(No. 2):304-310.

Thomson, J.
- Civil Engineer, Nashville, TN.
- (M)AJSA, 1833, *23*(No. 1):107-113.
- (M)AJSA, 1833, *24*(No. 1):73-77.

Tiarks, J.L.
- Astronomer.
- (M)AJSA, 1829, *15*(No. 1):41-53.

Twining, Alexander C. (1801-1884)
- Graduated from Yale in 1820.
- Civil Engineer in Connecticut.
- Professor of mathematics, civil engineering and astronomy at Middlebury College (1839-1849).
- (P)AJSA, 1825, *9*:86-90.
- (M)AJSA, 1826, *11*(No. 1):184-188.
- (M)AJSA, 1834, *26*(No. 2):320-351.

Wallace, J.
- Professor from Columbia, South Carolina.
- (P)AJSA, 1824, *7*:278-285.
- (R)AJSA, 1825, *9*:93-103.

Webber, Samuel (1759-1810)
- Graduated from Harvard in 1784.

- Hollis Professor of Mathematics and Natural Philosophy at Harvard, 1789-1804.
- President of Harvard, 1806-1810.
- Wrote influential textbook *A System of Mathematics*.
- (M)AAAS, 1793, *2*:20-21.

West, Benjamin (1730-1813)

- Rhode Island almanac-maker.
- Professor of mathematics and natural philosophy at Brown (1786-1799).
- (M)APS, 1771, *1*:97-104.
- (M)AAAS, 1785, *1*:156-158.
- (P)AAAS, 1785, *1*:165-172.

Wheeler, A.D.

- Instructor of Latin Grammar School in Salem, MA.
- (P)AJSA, 1831, *20*(No. 2):295-296.
- (P)AJSA, 1834, *25*(No. 1):87-89.

Wilder, C.

- New Orleans resident.
- (P)AJSA, 1829, *16*(No. 2):271-282.
- (P)AJSA, 1830, *18*(No. 1):38-46.
- (P)AJSA, 1830, *18*(No. 2):276-277.
- (P)AJSA, 1831, *20*(No. 2):285-290.

Wilkie, Rev. Daniel

- (P)AJSA, 1833, *24*(No. 1):68-69.

Willard, Joseph (1738-1804)

- Graduated from Harvard in 1765.
- Helped to form the AAAS.
- President of Harvard, 1781-1784.
- (M)AAAS, 1785, *1*:1-61.
- (M)AAAS, 1785, *1*:70-80.
- (M)AAAS, 1785, *1*:129-142.
- (M)AAAS, 1785, *1*:151-155.
- (M)AAAS, 1785, *1*:318-321.

Williams, Rev. Samuel (1743-1817)

- Graduated from Harvard in 1761.
- Hollis Professor of Mathematics and Natural Philosophy at Harvard, 1780-1788.
- Lectured on astronomy and natural philosophy at the University of Vermont while a minister in Vermont in the first decade of the 1800s.
- (M)AAAS, 1785, *1*:62-69.
- (M)AAAS, 1785, *1*:81-123.
- (O)AAAS, 1785, *1*:234-245.
- (O)AAAS, 1785, *1*:260-311.

- (O)APS, 1786, *2*:118-122.
- (M)APS, 1786, *2*:246-250.
- (M)AAAS, 1793, *2*:22-36
- (O)APS, 1793, *3*:115.

Williamson, Hugh (1735-1819)

- Graduated from College of Philadelphia in 1757.
- Pennsylvania physician.
- Professor of mathematics at College of Philadelphia (1760-1763).
- (M)APS, 1771, *1*:Appendix(27-36).

Winthrop, James (1752-1821)

- Son of famous Harvard professor John Winthrop.
- Graduated from Harvard in 1769.
- Considered for Hollis chair to replace his father, but rejected. Served as Harvard librarian.
- Published several articles in *AAAS* including failed attempts to trisect and angle and duplicate a cube.
- (M)AAAS, 1785, *1*:159-161.
- (P)AAAS, 1793, *2*:9-13.
- (P)AAAS, 1793, *2*:14-17.
- (O)AAAS, 1793, *2*:127-130.
- (M)AAAS, 1804, *2*(part 2):20-22.

Wright, Elizur (1762-1845)

- Graduated from Yale in 1781.
- Professor at Western Reserve College, Ohio (1829-1833).
- Actuary in Connecticut and Massachusetts.
- (M)AAAS, 1804, *2*(part 2):25-39.
- (P)AJSA, 1828, *14*(No. 2):330-350.
- (R)AJSA, 1829, *16*(No. 1):53-59.
- (P)AJSA, 1832, *22*(No. 1):74-82.
- (P)AJSA, 1833, *24*(No. 2):298-311.
- (P)AJSA, 1834, *25*(No. 1):93-103.

Appendix D

John Farrar's *Cambridge Course of Mathematics*

Textbook	Original Author	First Edition	Final Edition	Number of Editions
Algebra	Euler	1818	1836	4
Arithmetic	Lacroix	1818	1834	4
Algebra	Lacroix	1818	1837	5
Geometry	Legendre	1819	1841	9
Trigonometry	Lacroix and Bézout	1820	1837	4
Trigonometry	Malortie, Puissant, Bonnycastle, Delambre, and Bézout	1822	1840	4
Calculus	Bézout	1824	1836	2

www.ingramcontent.com/pod-product-compliance
Lightning Source LLC
Chambersburg PA
CBHW071945110426
42744CB00030B/294